THE EXPLODING SUNS

THE EXPLODING

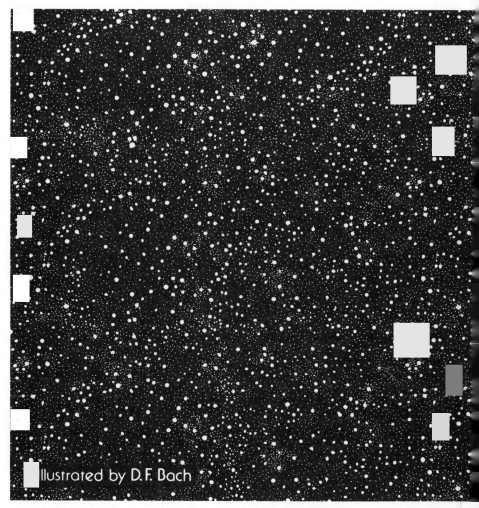

Illustrated by D. F. Bach

[T·T] TRUMAN TALLEY BOOKS / E. P. DUTTON / NEW YORK

SUNS

The Secrets
of the Supernovas

ISAAC ASIMOV

Published in the United States by
Truman Talley Books · E. P. Dutton,
2 Park Avenue, New York, N.Y. 10016

Library of Congress Cataloging in Publication Data
Asimov, Isaac, 1920-
    The exploding suns.
    "A Truman Talley book."
    1. Stars—Popular works. 2. Supernovae—Popular
works. 3. Cosmology—Popular works. I. Title.
QB801.6.A85   1985   523.1   84-21077

ISBN:0-525-24323-2

Published simultaneously in Canada
by Fitzhenry & Whiteside, Ltd., Toronto

Designed by Mark O'Connor

10 9 8 7 6 5 4 3 2

Dedicated to
the memory of
Marvin Grosswirth
(1932–1984)

CONTENTS

Contents

THE
EXPLODING
SUNS

NEW STARS

The Unchanging Sky

If we look at the sky on a clear, moonless night, we are bound to be impressed by the quiet changelessness of it all. The stars shine in fixed patterns with fixed brightness. They seem to move in a steady, unvarying circle having its center near the North Star (assuming we are viewing the sky from the northern hemisphere) and completing a rotation in twenty-four hours.

Each night the view at exactly midnight shifts slightly, as though the Sun were turning against the background of stars, but much more slowly than that of the daily movement. The Sun completes this slower turn in 365¼ days. Both turnings, however, are regular, and the pattern of the stars does not change while the turning takes place.

The Greek philosopher Aristotle (384–322 B.C.) felt this changelessness of the sky to be a law of nature. On Earth, he believed, all things changed and decayed, were first formed and then destroyed, while all things in the sky were changeless, perfect, and permanent. Things on Earth tended to be at rest (if not alive) or to fall, while things in the sky never rested but moved in stately, never-ending circles.

Aristotle even considered Earth and sky to be fundamentally different in constitution. Everything on Earth was made up of four "elements," four basic varieties of material—earth, water, air, and fire. The sky and all the objects in it, however, were made up of a fifth element, perfect and naturally glowing, which he called *aither*, from the Greek word for "blazing." (It is more familiar to us in the Latin spelling of "aether.")

There were undoubtedly many other early thinkers who believed in the changelessness of the heavens, but Aristotle was the most eminent—it is his works that have survived, so it is he who has always been taken as the great authority for this view.

It is a reasonable view because, for one thing, it fits in with the common observations we all make. Each of us can see with our own eyes that things on Earth come into being, grow, change, deteriorate, decay, and come to an end. On the other hand, the Sun and all the other heavenly bodies seem to go on forever with no change at all.

To be sure, there are some phenomena that seem to argue against Aristotle's thesis of heavenly changelessness, and if we observe carefully, we will note them. There *are* changes in the heavens, even quite obvious ones. Clouds come and go, for instance, thicken to a solid overcast, or thin out to nothing. Rain and other forms of

2

precipitation seem to fall on Earth out of the sky and then, eventually, cease.

Clouds and precipitation, however, are manifestations that exist in the air, and air is one of the four Aristotelian elements that are part of the Earth. Aristotle thought so, and modern astronomers certainly agree with him in this. Aristotle considered the atmosphere to extend up to the Moon, which was the nearest to us of all the heavenly bodies. The aether of the sky and the property of changelessness began with the Moon, he felt, and included everything beyond it and nothing below it.

There are other changes in the sky, however, besides the weather. Sometimes, when watching the quiet sky of nighttime, you may become aware of a point of light moving across the black vault, dimming as it does so, and soon fading out. It seems for all the world as though a star has come loose from the sky, slid rapidly across it, and, perhaps, fell to Earth. We sometimes call it a "shooting star," but it isn't really a star; for no matter how many of them we see, no star is ever missing from the vault of heaven as a result.

To Aristotle the shooting stars were also phenomena inside Earth's airy envelope, inside its atmosphere. We call them *meteors*, therefore, from a Greek word meaning "things in the air." The term refers, properly, only to the streak of light, and in that respect Aristotle was right, for that streak appears in the atmosphere. It is caused by some small object, varying in size from a boulder to a pinhead, that moves through space and happens to collide with the Earth. In passing through Earth's atmosphere at a very rapid speed, air resistance warms it to a white-hot glow.

The objects themselves are now called *meteoroids*. The smaller ones vaporize entirely long before they reach the surface of the Earth and float downward very slowly as a fine dust. The large ones survive the ordeal, in part at least, and a fragment or fragments can strike the Earth. The surviving fragments are *meteorites*. (Scientists were reluctant to accept the fact that solid objects could fall from the sky until the early 1800s.)

Then, too, various comets also appear and disappear irregularly in the sky, having odd and irregular (and therefore imperfect) shapes. Sometimes they change in appearance from night to night. To Aristotle, however, comets were regions of flaming vapors in the upper atmosphere and were therefore part of Earth, not of the sky. (Here he was flatly wrong, but his error was not demonstrated until the late 1500s.)

If we eliminate the weather, the meteors, and the comets, all that is left to consider is the Moon and the heavenly objects that lie beyond.

The Moon itself certainly exhibits changes. It changes shape from night to night, taking on a succession of *phases*, from a Greek word meaning "appearance." Even when the Moon is full and is a smooth circle of light (thus displaying the perfection of form one would expect of a heavenly body), there are shadows and blotches upon it that are clearly imperfections.

There were two ways of explaining this away. A number of people in ancient and medieval times pointed out that since, of all heavenly bodies, the Moon was nearest to Earth, it was also most exposed to the influence of the imperfect and "corrupt" Earth. The blotches, therefore, were Earthly exhalations.

Another rationalization of the Moon's changes was to argue that change can be tolerated in a perfect heaven if the change is cyclic, repeating itself over and over unendingly. Again, irregularity was not necessarily imperfect if the irregularity never changed.

Thus, the Moon's blotches never changed, and its

4

phases repeated themselves in so regular a fashion that it was simple to predict what the Moon's phase would be on any night for years ahead.

Another questionable point about the Moon was this: While it rose in the east, moved westward across the sky, and set in the west, as did the Sun and the stars, it did not exactly accompany the stars. Each night, the Moon was to be found in a different portion of the heavens relative to the starry background, and closer observation showed that it moved steadily west to east against that background, making a complete circle of the sky in a little over twenty-seven days.

The Sun, too, moved west to east against the starry background, as I mentioned earlier. The Sun's motion is considerably slower than the Moon's, for it takes 365¼ days to make a circuit of the sky.

The motions of the Moon and Sun against the background of stars are not *quite* regular, but worse yet, in the eyes of the ancients, was the case of five of the brightest stars, which were also seen to move against the starry background. These were given the names of gods by the awed observers, and we still use the divine names given them by the Romans. We know the five as Mercury, Venus, Mars, Jupiter, and Saturn. They do not move steadily west to east against the starry background as do the Moon and the Sun. They would, instead, occasionally slow, then turn and move in "retrograde" fashion from east to west. They would then turn again and, for a period of time, move in the ordinary fashion, repeating the process over and over again. They would indulge in retrograde movement anywhere from once a year or so (as in the case of Mars) to twenty-nine times a year (as in the case of Saturn).

The Greeks called the seven objects—the Moon, the Sun, Mercury, Venus, Mars, Jupiter, and Saturn—*plan-*

5

The ancients had no idea of the enormous size of the Sun, even after they had worked out the size of the Moon.

etes ("wanderers") because they wandered among the stars. That word has come down to us as *planets*.

In order to explain the separate motions of the planets, the Greeks supposed that each of the planets was fixed to a separate sphere surrounding the Earth, with these spheres nested one within the other. Judging that the more rapidly a planet moves across the sky the closer it is to Earth, the Moon was considered as being embedded in the innermost sphere, Mercury in the next, and then, in order, Venus, the Sun, Mars, Jupiter, and Saturn. Each sphere was absolutely transparent ("crystalline") and could not be seen. (It is because of these spheres that we speak of "heavens" in the plural even today.) Each sphere was supposed to turn, and this turning was responsible for the planetary motion across the sky.

Plato (427–347 B.C), who was Aristotle's teacher, held that only regular circular movements were perfect. To account for the irregular motions, the planets would have to be pictured as moving in combinations of regular circular movements if the heavens were to be considered perfect. Aristotle, and the Greek thinkers who followed him, tried to work out ever more complicated combinations of circular movements that would suffice to make the planets move in the irregular way they were observed to move, and yet keep them from seeming imperfect.

Nowadays, we know that the meteoroids, the comets, and the seven planets are all part of what we call the "solar system," as is the Earth itself. The various members of the solar system (including the Earth) circle about the Sun, which was called *sol* by the Romans, and which is why we speak of the solar system. The Sun is a star that looks different from other stars only because it is so close to us.

If we ignore the solar system and consider only the stars beyond it, then Aristotle's notion of the changelessness of the heavens appears to be correct. We can watch the stars from night to night and year to year with our unaided eyes (as the ancients watched) and we would very likely see no change.

Change Among the Stars

To the ancients (and to our eyes, too, if we didn't know better), the stars, some 6,000 of them, seemed to be fixed to the outermost sphere, the one that lay beyond the sphere of Saturn. (The stars were therefore termed "fixed stars" to differentiate them from the "wandering stars," or planets, that moved independently of that sphere.)

The outermost sphere of stars was not transparent but was black, and the stars shone against it like very tiny luminous beads. The whole black sky turned with complete regularity once a day, carrying the stars with it so that they did not change position relative to each other. When the Sun rose, the sky turned blue and the stars vanished, but that was only because the blaze of the Sun drowned them out.

Surely, Aristotle's notion of the perfection of the heavens seemed to apply to the fixed stars without any disturbing questions.

That, however, brings us to the Greek astronomer Hipparchus (190-120 B.C), the greatest of all the Greek astronomers. In fact, considering that he had no instruments to work with, save for some very simple ones he had invented himself, and the very limited records of previous astronomers, what he accomplished is enough to mark him as among the greatest of all astronomers.

Hipparchus worked on the island of Rhodes off the southwestern coast of what is now Turkey and, to explain the apparent motion of the planets, he devised a better system of circle-combinations than anyone else had managed to do in the two centuries after Plato's death. Hipparchus' system survived, with minor improvements, for 1,700 years.

A later astronomer, Claudius Ptolemaeus (A.D. 100-170), who lived about three centuries after Hipparchus, summarized the Hipparchean system, with some improvements, about A.D. 150, in a book that survived into modern times, although none of Hipparchus' writings did.

As a result, we refer to the astronomical system in which Earth is pictured as at the center of the universe, with all other astronomical bodies circling it, as the "Ptolemaic system," which is rather unfair to the earlier astronomer.

In 134 B.C., Hipparchus prepared a star catalog, the first good one ever prepared. In it, he listed 850 of the brighter stars. (Ptolemy incorporated the chart into his book and added 170 more.) Hipparchus located each star according to a system of latitude and longitude and gave its brightness according to a system of "magnitude" he invented. By this system, the stars were divided into six classes. The "first magnitude" included the twenty brightest stars in the sky, and the "sixth magnitude" included the 2,000 or so stars that are just barely visible on a moonless night to people with sharp eyesight. The second, third, fourth, and fifth magnitudes lay between these extremes.

It is rather surprising that Hipparchus bothered to do this at all. The stars were not considered important to ancient astronomers. They were simply a speckling of background against which the planets moved. It was the planets that were important and that occupied virtually all the attention of the early astronomers. Most people thought that the planets in their motion exerted an influence on Earth and on human beings, and, if a system could be worked out that could predict their motions exactly, it might also be possible to work out their influences on the future fate of each individual. The development of such a system of *astrology*—planet readings—was of devouring interest to everyone in ancient times.

The Sun, the Moon, and the five star-like planets all moved along a narrow strip of the sky that was divided into twelve regions, each occupied by a particular grouping of stars in which the imaginative ancients marked out the figure of some object, usually an animal. Each grouping is called a *constellation*, and the twelve constellations through which the planets moved are called the *zodiac*, from the Greek word meaning "circle of animals."

Why twelve constellations in the zodiac, by the way? Because the Sun remained in each constellation for one month, that is, for one complete circle of the zodiac by the Moon.

Eventually, astronomers divided the rest of the sky into constellations also. In modern times, when astronomers traveled southward and could study the stars in the far south (stars that can never be seen from the northern latitudes, where most of the ancient civilizations were located), that portion was divided into constellations as well. Nowadays, there are eighty-eight constellations dividing the entire sphere of the sky, but still it is the twelve constellations of the zodiac that remain most interesting to some credulous human beings.

Hipparchus, watching the sky from night to night and following the position of the planets so that he might work out his system of planetary motion, had to observe the fixed stars that appeared adjacent to the planets. He must, very likely, have memorized the positions of all the brighter stars in the sky and, especially, the stars of the constellations of the zodiac.

According to the Roman scholar Pliny (A.D. 23-79), who wrote an encyclopedia of human knowledge two centuries after Hipparchus, the star catalog the astronomer had prepared was inspired by a "new star" that had appeared in Scorpio, one of the constellations of the zodiac.

We can imagine Hipparchus' astonishment at observing, one night, the appearance of a star that had not been there the night before.

Astonishing? Impossible! How could there be a new star in a changeless perfect heaven?

Hipparchus gets the idea—so obvious in hindsight—of mapping the stars.

He must have studied that new star night after night with incredulity and have seen it gradually fade until, finally, it disappeared.

It might have occurred to him that this was not necessarily a unique phenomenon. Perhaps new stars appeared repeatedly and then faded away, and it may be that this went unnoticed because people didn't study the stars very closely and couldn't tell that something new had appeared. Even astronomers might not be sure whether an object were really new, so that the star would not be properly studied and would fade away with no one having taken note of it.

By preparing a star chart that included the true permanent stars, Hipparchus made it that much easier for other and later astronomers to recognize any occasional new star that would appear. A suspicious object need merely be checked against the chart. That alone would make a star chart worthwhile.

As a tale, Hipparchus and his new star may be interesting, but is it true? Pliny, the source of the story, was a prolific writer who had very little discrimination. He tended to report everything he heard, so we don't know how reliable his source was. Did he find it in one of Hipparchus' own writings, for instance, which may still have existed at that time? In that case we could take it as reliable. On the other hand, he might simply have been dealing with a vague third-hand report that he found interesting.

The next person to refer to Hipparchus' new star was a Roman historian of the 200s. Two centuries after Pliny, he referred to Hipparchus' new star as a *comet*.

That may mean nothing. Any unrecognizable object in the sky in those days might have been referred to as a comet (as today it might be called a UFO).

It remains true, though, that in all the surviving records of Greek and Babylonian astronomy there is no mention of any new star, no mention of any temporary star that appeared in the sky where no star should be, except for this one vague tale concerning Hipparchus.

Nowadays, we know very well that new stars *do* appear, and fairly frequently at that—some of them are even quite bright. Why, then, were they not reported in ancient and medieval times?

As I said, new stars are hard to recognize. Any casual watcher of the sky merely sees a large number of stars higgledy-piggledy. Add a new star on some particular night, even a quite bright one, and no one but a dedicated astronomer is at all likely to notice. Even astronomers might not notice. The astronomers of ancient Babylonia and Greece were, for the most part, watching the planets and those stars of the zodiac that were in the immediate vicinity of planetary positions. They might well miss a new star outside the zodiac. Even Hipparchus might have noticed this new star only because it was in one of the constellations of the zodiac.

Then, after Aristotle's notion of the perfection of the heavens caught on, that would have introduced another barrier. Once astronomers got the fixed idea that there was no change in the heavens, they would grow very reluctant to report a change. They would fear being disbelieved and worry that their reputation might suffer. Indeed, they would very likely mumble to themselves that their eyesight was going and that they were suffering an optical illusion. In this way, they would avoid the risks of an unpopular announcement.

Eventually, to report such a change might even have come to involve sacrilege. The medieval astronomers, whether Christian or Muslim, saw in the perfection of the heavens (particularly of the Sun) a symbol of the perfection of God. To find any flaw in that perfection would be to cast doubt on the workmanship of God, and that would be a most serious thing to do. It might well seem to them that even Earth was imperfect only because of the fact that Adam and Eve had eaten the forbidden fruit in the Garden of Eden. Had they not done so, Earth might be as perfect as the heavens.

It may be, therefore, that throughout the early history of astronomy new stars did occasionally appear, and that astronomers either didn't notice them, or didn't believe their eyes, or just prudently kept their mouths shut.

13

China's "Guest Stars"

Europe and the Middle East, however, were not the only homes of civilization.

For a 2,000-year period, between 500 B.C. and A.D. 1500, China was far ahead of the West in science and technology. Throughout ancient and medieval times, Chinese astronomers kept a close watch on the sky and recorded anything unusual that happened anywhere. They were not hampered by dogmatic beliefs of perfection, and theirs was a relatively secular society in which fear of supernatural beings did not unduly restrict their thinking.

For instance, they did report a comet in the sky in 134 B.C., and that supports the Roman historian's account of what it might have been that Hipparchus had seen.

To be sure, the Chinese didn't study the sky for purely intellectual reasons. They, too, like the Babylonians and Greeks, were interested in astrology. They had worked out meanings for anything at all that might happen in the sky and used them to proclaim the likelihood of various future events on Earth.

Since the events, as proclaimed by heavenly omens, were often disastrous—with astronomical observations seeming to presage wars, plagues, and death—it was necessary for the realm, for various high nobles epecially, and even for the Emperor himself, to be prepared to take action that would avert or ameliorate the event. If something evil took place and there had been no warning, it would not have been unusual for the court astronomers to be executed.

Consequently, the Chinese astronomers watched very carefully, and, for one thing, they painstakingly recorded any "guest star" that temporarily took up residence among the permanent stars. More than fifty such

new stars were recorded in the annals, stars that were totally missed by Western astronomers. Korean and Japanese astronomers, who picked up Chinese science and
technology, also recorded some of them.

A few of the new stars recorded by the Chinese were
very bright and remained visible for six months or more.
Five such particularly bright new stars were reported in
ancient and medieval times. In A.D. 183, for instance, the
Chinese reported a very bright new star in the constellation of Centaurus, and in A.D. 393 a less bright one in
Scorpio.

It is not surprising that these were not reported in
Europe. In those centuries, Greek astronomy had declined
and grown extinct (there were no Greek astronomers of
importance after Ptolemy), and the Romans were never
interested in any phase of science.

The new star in Scorpio was probably no brighter
than Sirius (the brightest permanent star in the sky),
and unless there was someone who studied the sky pro-
fessionally, who happened to be looking in that direction,
and who had either memorized that part of the sky or had
a star chart to consult, it would not be at all surprising for
the star to have gone unnoticed.

Furthermore, although the new star in Scorpio remained visible for about eight months (according to the
Chinese), it remained as bright as Sirius for only a few
nights. Then it steadily faded, and the dimmer it got the
less likely it was to be noticed by anyone not as intent as
the Chinese astronomers.

The new star of 183 in Centaurus, according to the
Chinese reports, was much brighter than the one that was
to appear in Scorpio two centuries later. For some weeks,
the new star in Centaurus may have been brighter than
anything in the sky, except for the Sun and the Moon. It
would seem to have been impossible to miss, but it was

far in the southern sky, and that increased the difficulty of observation, even for a very bright object. From the Chinese observatory in Lo-yang, the new star was never seen higher than 3° above the southern horizon.

In Europe, it would have been completely invisible from any part of France, Germany, or Italy, and it would have been just at the horizon as viewed from Sicily or Athens. However, it would have been passably visible from the more southerly latitude of Alexandria, which was then the center of Greek science.

Nevertheless, it wasn't reported by Greek astronomers. At least if any Alexandrian had noted the brilliant star on the southern horizon, respect for the Aristotelian view kept him from reporting it; or, if he did report it, the ancient world of science simply didn't accept it, so the report does not survive.

For six centuries after the star of 393 of Scorpio, there was no new star of remarkable brightness in the Chinese records. Then, in 1006, there came the report of a new star in the constellation Lupus, which is adjacent to Centaurus and is therefore also far in the southern sky.

Despite its far southern location, it was reported by both Chinese and Japanese astronomers. In the West at that time, astronomy was best practiced by the Arabs, who were then at the height of their scientific preeminence. There are at least three reports from the Arabs as well.

It is not at all surprising that this new star was widely seen: the reports all agree on its brilliance. It is estimated by some modern astronomers to have been possibly 200 times as bright as Venus at its very brightest, and therefore perhaps one-tenth as bright as the full Moon. It stayed visible to the eye for perhaps three years, though it could not have been brighter than Venus for longer than a few weeks.

16

The new star was high enough in the southern sky to be seen from southern Europe, and we might imagine people in Italy, Spain, and southern France looking toward the southern sky at night in awe and wonder. They didn't; or at least there is no report of their having done so. The chronicles kept by two monasteries, one in Switzerland and one in Italy, seem to make reference that year to something that could be interpreted as a bright star, but that's all.

Since it appeared in 1006, one might suppose that Europeans would have taken it as a sign that the world was coming to an end, for some people at the time thought that such an ending might be expected a thousand years or so after the birth of Jesus. But even that fearful possibility didn't seem to bring on any notice.

Then in 1054 (on July 4, according to some calculations, in premature celebration of the day) there blazed out another new star, and this time it was in the constellation of Taurus, well north of the Equator. Unlike the far southern new stars of 185 and 1006, it was clearly visible over the entire northern hemisphere. What's more, it was in the zodiac where it couldn't fail to be noticed.

Further, it was not merely as bright as Sirius, as the new star of 393 (also in the zodiac) had been. The new star in Taurus was at least two or three times as bright as Venus at its brightest. For three weeks it remained bright enough to be seen by daylight (if one knew where to look), and at night it could cast a dim shadow (as Venus can, under favorable conditions). It remained visible to the eye for nearly two years, and it may have been brighter than any new star in historic times, except for the one of 1006.

Later, it was thought that *only* the Chinese and Japanese astronomers reported this spectacular and easily visible object in the sky. There seemed to be no reports at all from either Europeans or Arabs.

How could that be? During the month of July 1054, when the new star was brightest, it must have been very conspicuous in the hours before dawn. Perhaps most Europeans were asleep at the time, or perhaps there was considerable cloudiness then. Or if the star were visible, perhaps those few who were awake and looking merely dismissed it as Venus, and possibly those who thought "But that can't be Venus" thought also of Aristotle and the perfection of God's handiwork and then uneasily looked away.

In the last few years, however, an Arab account has been discovered that seems to refer to a bright new star in 1054, and there is even an Italian manuscript that also seems to refer to it.

This is a great relief. There might be a feeling among those of us of the Western tradition that if there had been no accounts from Europe then the star could not really have existed. It might be easier to believe that some far-distant foreigners were fantasizing than that Europeans could not see what was before their eyes. However, as I shall explain later, even if there had been no reports at all from the West, we still have firm evidence that the Chinese and Japanese astronomers were absolutely correct.

In 1181, there was another new star reported by the Chinese and Japanese, this time in the constellation of Cassiopeia. This would make the new star clearly visible all across the northern hemisphere. However, it grew only as bright as the star Vega, the second brightest in the northern skies, and it went unnoticed in Europe.

Then, for four centuries, no new stars were observed. By the time the next new star appeared, conditions had changed. Chinese and Japanese astronomers were as skilled as ever, but there had been a rebirth in Europe. Now it was European science that led the world.

The First Nova

In 1543, the Polish astronomer Nicholas Copernicus (1473-1543) published a book that described the mathematics necessary to predict the position of the planets if one were to assume that the Earth, along with Mercury, Venus, Mars, Jupiter, and Saturn, all revolved about the Sun. (The Moon was still pictured as revolving about the Earth.) This assumption simplified matters considerably and led to better planetary tables, even though Copernicus still took the position that the planets moved in combinations of circular orbits.

The book, published at the end of Copernicus' life (a freshly printed copy was supposed to have been handed to him on his deathbed), aroused intense controversy. Very few people were willing to believe that the vast and heavy Earth was flying through space at an enormous speed, since there was no sensation of motion at all. It was at least a half century before astronomers accepted this "heliocentric" theory, although, in the interval, the view of the heavens as worked out by Hipparchus and Ptolemy was seriously shaken.

Three years after Copernicus' book was published, Tycho Brahe (1546-1601) was born in the southernmost province of Sweden, which was then part of Denmark. In early life he studied law, but when he was fourteen years old he observed an eclipse of the Sun, and this turned his attention to astronomy (fortunately for both him and astronomy).

His opportunity came in 1572, when he was twenty-six years old and was still generally unheard of in Europe.

Until that year, Europeans, including the astronomers, knew nothing of new stars. There was the vague tale of Hipparchus' new star, which could easily be dismissed as just an ancient fable, since Ptolemy said nothing about it. The few mentions in one or two Western chronicles of the new stars of 1006 and 1054 were so obscure that no astronomer of the 1500s could possibly have known about them.

And that was it! Certainly, no European astronomer knew anything about the records compiled by the Chinese, the Koreans, and the Japanese.

Then, on November 11, 1572, Tycho Brahe, walking out of his uncle's chemical laboratory, saw a new star, one he had never seen before. It was in Cassiopeia, high in the sky, and brighter than any of the stars in that well-known constellation. It could not be missed by anyone as aware of the map of the sky as Tycho was.

Like the new star of 1054, the new star in Cassiopeia was considerably brighter than Venus at its brightest. Nor could it be mistaken for Venus by any astronomer, for it was far outside the belt of the constellations of the zodiac and far from any place that was ever occupied by any planet.

In great excitement, Tycho asked everyone he passed to observe the star in the hope that they could tell him if it had been there the night before.

Everyone he asked said they saw the star, so there was nothing wrong with Tycho's eyesight. None of them, however, could say whether it was a new star or not, or, if a new star, when it had first appeared. It was a bright star, but for all anyone else could say it might have been there every night of their lives.

Tycho, however, was convinced that nothing like it had been in the sky the last time he had looked. But since he had been engaged in chemical experiments in his uncle's laboratory, he had not been watching the sky for a while. He couldn't swear it had not been present the previous night or even the past several nights. (Interest-

Tycho is startled by a new star in the "changeless" heavens.

ingly, a German astronomer, Wolfgang Schuler, seems to have noticed the new star just before dawn on November 6, five days before Tycho had seen it.)

Tycho now did something no other astronomer had ever done before. He began a series of nightly observations. He had constructed some excellent instruments during an earlier stay in Germany, and he used one of them at once. This was a large "sextant" with which he measured, in angular units, the distance of the new star from the other stars in Cassiopeia. He calibrated his instruments carefully in order to correct any errors due to imperfection in their construction, and he made allowance for the refraction of light by the atmosphere (being the first astronomer ever to do so). He also kept careful records of every observation he made and all the conditions under which he made them.

He didn't have a telescope, to be sure, for the instrument was not to be invented for another thirty-six years, but he made a reputation as the best pre-telescopic observer in astronomical history. Indeed, it was his observations of the new star that, perhaps even more than Copernicus' new theory, marked the beginning of modern astronomy.

The new star was near enough to the North Star for it to turn about that star in such small circles that it never dipped below the horizon, remaining always in the sky. Tycho could therefore observe it at any hour of the night. And he was startled to discover that it shone so brightly that he could even see it during the day.

It remained bright for a relatively brief period, however; each night it faded. By December 1572, it was no brighter than Jupiter; by February 1573, it was barely visible, and by March 1574, it seemed to disappear. It had remained visible, under Tycho's observation, for 485 days. Chinese and Korean astronomers also noted the new star, but they did not make the accurate measurements of its position that Tycho did. They had begun to fall behind the Europeans.

What was the new star? Was it an atmospheric phenomenon, as it would have to be if Aristotle's belief that the heavens were perfect and unchangeable were true? Could an atmospheric phenomenon remain in place for 485 days—*exactly* in place, too, because Tycho's careful measurements could detect no measurable shift in its position relative to the other stars in the constellation during all that time.

Tycho even tried to determine its distance directly. This can be done by measuring the "parallax" of an astronomical body—that is, by noting the manner in which it shifts its apparent position relative to other, more distant, bodies when seen from different places.

The Moon, which is the nearest of the heavenly bodies, has a small parallax but one that is large enough to be measured without a telescope. Since the time of Hipparchus, its distance was known to be thirty times the diameter of the Earth, so that the Moon is about 380,000 kilometers (240,000 miles) from Earth in modern units of length.

Anything with a smaller parallax than that of the Moon had to be more distant than the Moon and had to be part of the heavens. The new star had a parallax so small that it could not be measured at all by Tycho's best efforts. It was *not* an atmospheric phenomenon, therefore, but was a star like other stars.

This was so important that Tycho, after considerable hesitation, decided to write a book about it. Tycho considered himself a nobleman, and noblemen did not stoop to explain, in those days, to lesser mortals. It was only the vital nature of his finding that convinced him he must do it.

The book, written in Latin, as was then the custom for scholarly books, appeared in 1573. It was a large-sized book, but not a long one, for it contained only fifty-two pages. It had a fairly long title, but it is almost always referred to by a shorter version, *De Nova Stella* ("Concerning the New Star").

The book contained much material about the astrological significance of the new star, for Tycho firmly believed in astrology, as did most astronomers of the time. Putting aside the astrology, Tycho described the new star's brightness and how it faded from week to week. He gave the measurements of its position and even made a drawing of the surrounding stars with the position of the new star shown so that people could visualize exactly what Tycho had seen.

Most important, he explained that its position had never changed, and that it had no measurable parallax. It was a star, a *new* star. The heavens had clearly undergone change.

The book caused a sensation, for it marked the end of Greek astronomy. All notions of the permanence and perfection of the heavens had to be abandoned. In fact, a bright comet appeared in 1577 that *did* move with respect to the stars, but Tycho showed that it, too, had no parallax, so it appeared that even comets were farther than the Moon and part of the heavens and were *not* atmospheric phenomena.

With the publication of his book, Tycho at once became the most famous astronomer in Europe. Further, the word *nova* (meaning "new") in the title of his book came to be used for the new star and for all new stars. From that day on, a new star in the heavens has been called a *nova*.

The plural of "nova" in Latin is "novae," and for much of the time since Tycho that plural has been used. However, we are growing less Latin-minded these days, and so the plural of "nova" is almost always given as "novas" now. (It hurts my pedantic soul a bit to do so, but I will use "novas" as the plural in this book.)

More Novas

One of the consequences of Tycho's nova was that many astronomers began to watch the stars more closely, instead of concentrating on the planets. The discovery of a nova, it was clear, might make one famous. Within a generation, then, it became clear that changes in the fixed stars were not all that rare.

In 1596, the German astronomer David Fabricius (1564-1617), a friend of Tycho, located a star in the constellation of Cetus that had not been there before. It was of the third magnitude, meaning that it was a star of only middling brightness. But astronomers were no longer going to let anything get past them.

Was it actually a new star? There was no difficulty in coming to a decision, for one had only to continue to observe it. In time, the new star dimmed and disappeared, and Fabricius felt himself justified in announcing that he had indeed discovered a nova.

The next nova involved the German astronomer Johannes Kepler (1571-1630).

Kepler had worked with Tycho in the last years of the older man's life. Tycho, having spent many years making careful measurements of the changing positions of Mars against the starry background, hoped that he would be able to use those observations to demonstrate the truth of a compromise position he had conceived concerning the planetary orbits. He wanted to show that Mercury, Venus, Mars, Jupiter, and Saturn all revolved about the Sun; while the Sun, with these planets circling it, revolved about Earth.

When Tycho died in 1601, he left all his records to Kepler in the hope that his assistant would use them to substantiate proof of the "Tychonic system."

Kepler could not, of course, confirm it. What he did confirm, in 1609, was that Mars did *not* move about the Sun in a circle or in a combination of circles as Plato had insisted and as all subsequent Western astronomers, including Copernicus, had supposed. Mars moved around the Sun, instead, in an elliptical orbit with the Sun at one focus. Kepler went on to show that all the planets moved in elliptical orbits.

In doing this, Kepler had finally worked out the actual description of the solar system. His system, and not that of Copernicus, fit reality. In the nearly four centuries since, astronomers have made no substantial improvement on Kepler. More inclusive theories have been worked out and new planets have been discovered, but the elliptical orbits remain and, it seems quite certain, will continue to remain.

In 1604, however, before Kepler had fully developed his system, a new star blazed in the constellation of Ophiuchus. It was brighter than Fabricius' nova, though by no means as bright as Tycho's. The nova of 1604 was about as bright as Jupiter, and was perhaps only one-fifth as bright as Venus at its brightest.

Still, this was a stunning occurrence in a sky that was now scanned with total absorption by various astronomers. Kepler, and Fabricius, too, made careful measurements of the position of the 1604 nova and of its changes from week to week. It took an entire year before it vanished.

Thus, between 1572 and 1604, a one-generational span of thirty-two years, three novas had been observed in the heavens, two of them quite bright. All three were spectacular phenomena, though not as rare as their observers might have suspected.

2

CHANGING STARS

Seeing the Invisible

In 1604, at the time of Kepler's nova, man's scan of the stars remained much as it had always been. The sky still seemed to be a sphere made of some solid substance. The stars were luminous beads on the firmament and seemed fixed there.

Occasionally, a tiny bright irruption—a nova—was placed against the firmament through some unknown agency. These new luminous marks flashed but always faded away. The brighter they shone, the longer they took to wink out, but sooner or later they all disappeared.

Once a nova faded away, might it continue to exist but merely be too dim for the human eye to discern? For that matter, might there be stars that were *always* too dim to see? Might there be stars that had existed ever since the universe had begun but had, from whatever start, been too dim to see and had therefore never been seen?

Some scholars must have thought so. A German cleric, Nicholas of Cusa (1401–1464), believed there to be an infinite number of stars spread through infinite space; that all the stars were actually suns, but seemed to be only faint dots of light (when they could be seen at all) only because they were at enormous distances; that around all the stars were planets, at least some of which were inhabited by intelligent beings. And, if there were an infinite number of stars and man saw only a few thousand, the vast majority of stars had to be too dim to see.

Nicholas' views sound very modern, but we haven't the faintest idea how he came by such notions. Nor could he persuade others of his startling ideas, since he had no observational evidence of any kind to support them.

An Italian scholar, Giordano Bruno (1548–1600), adopted Nicholas' notions, a century and a half later. By Bruno's time, however, the Protestant Reformation had taken place, churchpeople all over Europe had grown suspicious and insecure, and it had become much more dangerous to espouse strange-sounding ideas. In addition, Bruno was a strongly opinionated man who seemed to enjoy shocking and offending people. In the end, he was burned at the stake.

Bruno had no evidence for his notions, either. At the time of his death, virtually no one believed in stars that were too dim to see. After all, why should such invisible stars exist? Why would God have created them? To some, it was sacrilegious to suggest that God would create anything so useless.

In 1609, another Italian scholar, Galileo Galilei

(1564-1642), heard that a tube with lenses at each end had been invented in the Netherlands, one that could make objects appear larger and closer. He began experimenting at once and in no time had what we now call a telescope. With it, Galileo did something new and daring—he turned it on the sky.

Galileo's telescope was a small and primitive affair, but it was the first time anyone had ever scanned the night sky with something more than the unaided eye. The telescope gathered more light than the eye itself could, and it focused that increased quantity of light on the retina. As a result, everything looked larger or brighter or both. The Moon looked larger and showed more detail. So did the Sun, if one took precautions against being blinded by it. The planets looked larger and became little circles of light. The stars were so small that even if they became larger they still were not large enough to appear as more than mere points of light—but at least those points of light became brighter.

With his telescope, Galileo saw new and astonishing things everywhere he looked. He saw mountains and craters on the Moon, as well as flat areas he thought were "seas." He saw spots on the Sun. He saw four satellites circling Jupiter. He saw that Venus showed phases like the Moon. From what the telescope showed, it seemed very likely that the planets were worlds, as Earth was, and perhaps just as changing and imperfect. Even the Sun, with its new-found spots, was clearly imperfect. As for Venus, its phases couldn't exist in the Ptolemaic system, in the manner Galileo had observed, but they could in the Copernican system.

Galileo's telescope immeasurably strengthened the Copernican view of the solar system, and this got him into trouble with the Inquisition, which forced him to deny the Copernican view. But that did the conservative

Galileo sees what no man has ever seen before, thanks to the small telescope he had devised.

religious forces no good, since all of scholarly Europe quickly accepted, first, Copernicus' contention that the Sun was at the center of the planetary system, and second, Kepler's ellipses.

And yet the very first discovery made by Galileo with his telescope had nothing to do with the solar system. When he looked at the sky for the first time, he turned his telescope on the Milky Way and found that it was not simply a luminous mist but that it consisted of unbelievable crowds of stars that could not be seen individually with the unaided eye. No matter where he looked in the sky, his telescope invariably showed him many more stars than his eye alone could see.

It was clear that there were vast numbers of stars too dim to pick out with the unaided eye, but that could be seen once they were brightened sufficiently by the telescope.

It followed, then, that when a nova dimmed and disappeared, it might not really disappear at all. It might merely grow too dim to observe with the unaided eye. In fact, a nova might not really be a new star at all—it might simply be a star that was too dim to see ordinarily and that suddenly brightened, became visible, then dimmed again, and retreated into invisibility.

In 1638, a Dutch astronomer, Holwarda of Franeker (1618-1651), sighted a star precisely in the sky region where Fabricius had seen his nova forty-two years earlier. Holwarda watched it fade, apparently disappear, then return. It turned out to increase and decrease in brightness every eleven months or so, and with a telescope it could still be seen even at its dimmest. At its low point it was at the ninth magnitude (allowing the least visible stars to be of the sixth magnitude, and continuing Hipparchus' system down into the least-bright levels reached by telescopes).

At its strongest, Fabricius' star was about 250 times as bright as it was at its weakest. It was not a nova in the strict sense, not a "new" star. Even so, it served to demolish the idea of the immutability of the heavens. A changing star, one that altered its brightness, was as much against Aristotle's dictum of permanence as a nova.

Stars of varying brightness are now called "variable stars," and Holwarda was the first to identify one. Nevertheless, variable stars that brightened suddenly and unexpectedly and failed to do so on a regular basis were still called "novas," even though the word meant "new." Fabricius' star, however, which *did* brighten and dim on a regular basis, was no longer considered a nova; it was merely a variable star.

The German astronomer Johann Bayer (1572-1625) invented a system, in 1603, of naming each star by a Greek letter and the name of the constellation in which it was located. He had given Fabricius' star the name of "Omicron Ceti" when he recorded its position during one of its periods of visibility. (He had failed to realize that it was the "nova" reported by Fabricius.) Once its variable nature was determined, the German astronomer Johannes Hevelius (1611-1687) named it "Mira," which is Latin for "wonderful."

Mira may have been wonderful because variability seemed such a strange and unusual property when it was first discovered, but variability did not remain strange and unusual for long. Before the end of the seventeenth century, three more variable stars were discovered. One of these was a very well known star, Algol, the second brightest star in the constellation Perseus, so that it is sometimes known as "Beta Persei."

In 1667, an Italian astronomer, Geminiano Montanari (1633-1687), noticed that Algol varied in brightness. It was no Mira, for the variation wasn't extreme. Algol was at a magnitude of 2.2 at its brightest and 3.5 at its dimmest, so that it shone about three times brighter at its brightest than at its dimmest.

This may have been noticed earlier by the Arabs. The mythical hero Perseus is usually represented as holding the head of the monstrous Medusa, whom he has just killed. The head of Medusa, who was so horrible a monster that the mere sight of her would turn men into stone, was represented by the star Algol. This name was given it by the Arabs, and it means the "ghoul" (or "demon"). Was this because it represented Medusa, or was it because the star varied in brightness and, in that respect, seemed to be defying the hallowed notion that the heavens were unchanging? Indeed, did the Greeks themselves uneasily note the variation and make the star represent Medusa for that reason?

In 1782, a seventeen-year-old English deaf-mute, John Goodricke (1764-1786), studied Algol carefully and discovered that the variation was absolutely regular. It went through one cycle of brightening and dimming in just sixty-nine hours. Goodricke suggested that Algol was a double star, one much dimmer than the other. The two circled each other, and every sixty-nine hours the dimmer one moved in front of its brighter companion so that the light of Algol temporarily faded. He turned out to be right, and there are about 200 such "eclipsing variables" now known.

Algol is not a true variable, therefore, for each of the pair of stars shines with perfect constancy and would not seem to vary at all if one star did not periodically get in the way of the other.

In 1784, Goodricke discovered that a star in the constellation Cepheus, one known as "Delta Cephei," is also variable. It is even less variable than Algol, being only twice as bright at its brightest as at its dimmest. Delta Cephei has also a very regular period, brightening and dimming every 5⅓ days. The manner in which its brightness rises and falls, however, is not the kind that can be easily explained by means of an eclipse. It dims more slowly than it brightens while an eclipsing variable should dim and brighten at the same rate.

Many other variable stars were discovered in the next two centuries, with their curves of brightening and dimming similar to that of Delta Cephei, though with periods ranging anywhere from two to forty-five days.

These are called "Cepheid variables." It was not until the 1920s that the English astronomer Arthur Stanley Eddington (1882-1944) was able to show that the curve could be explained by assuming the star to be pulsating—that is, regularly swelling in size and then contracting.

Most variable stars are such "pulsating variables"; some are short-period, some long-period, some regular, some irregular. Many thousands of all kinds are now known.

Novas are also listed among the variable stars, since their brightness varies with time. What makes them immediately different is that the variation is far greater than in the case of other variables. Novas brighten by tens of thousands of times, rather than by two or three times. They then fade off in a far more prolonged manner and to a far greater extent. What's more, other variable stars are cyclic, repeating the brightening and dimming over and over again at short intervals. Novas, however, are one-shots. If they do undergo repeated episodes of brightening, they do so at long and totally unpredictable intervals.

Movement and Distance

After the spectacular novas observed by Tycho and by Kepler and the full realization that the heavens changed, a century and a half passed without any new reports of novas. Indeed, Fabricius' star, which he had thought to be a nova, was shown not to be one after all.

This is not to say that no novas had irrupted. It was simply that any that had were not spectacular and had not been observed. Even with more and more sky watchers, there simply weren't enough astronomers to study every patch of night sky with sufficient intentness to recognize an unspectacular nova among the crowds of ordinary stars made visible by the new telescopes. Even today, when astronomers have at their disposal magnificent star charts together with advanced techniques of photography, novas may be missed at first and recognized only after having passed their initial peak. Novas may even go completely unnoticed until photographs taken much earlier are gone over in detail.

Nevertheless, the century and a half during which no novas were reported was not without important advances in the study of the stars.

Even after a hundred years of telescopic studies, it was still possible to believe that the sky was a solid sphere just outside the orbit of Saturn (which was the farthest known planet in 1700, as in ancient times) and that stars were little luminous points affixed to it. To be sure, the telescope had multiplied those points enormously, but the great firmament had room for all.

It was the English astronomer Edmond Halley (1656–1742) who first revealed that a comet traveled in a fixed orbit about the Sun and returned periodically. The comet he worked with has been called "Halley's Comet" ever since.

In later years, Halley worked on the problem of precisely positioning the various stars. As telescopes improved, so did accuracy.

Comparing his figures with earlier ones, Halley was astonished to note that the Greeks had apparently located some stars incorrectly. Even given the fact that the Greeks lacked telescopes, the stars' positions still seemed far wrong—especially since only a few of the brighter stars were misplaced.

Halley felt there was only one conclusion. The Greeks

were *not* wrong; it was the stars that had shifted position in the course of the previous sixteen centuries. In 1718, Halley announced that the bright stars, Sirius, Procyon, and Arcturus, had all moved noticeably since Greek times and slightly even since Tycho had recorded their latitude and longitude a century and a half earlier.

It seemed to Halley that the stars weren't fixed at all, but that they roved randomly through vast stretches of space like bees in a swarm. The stars were so far away, on the whole, that the distance they traveled was small indeed compared to their distance from Earth, so no motion at all could be detected from night to night or from year to year—until telescopes were sufficiently refined to make it possible to measure extremely small shifts in position.

If, however, positions were measured over generations and centuries, the shifts became noticeable, particularly among the nearer stars. Sirius, Procyon, and Arcturus must be among the nearer stars, he thought, which would account for both their brightness and their evident "proper motion."

But what were the distances of the stars? One could tell if one could determine the parallax of some of them. A nearer star would appear to change position relative to a much further one as the Earth orbited the Sun, shifting its own position from one side of the Sun to the other side—a shift of 300 million kilometers (186 million miles). The apparent motion of even the nearest stars in response was so small, however, that the telescopes of Halley's day, and even those of a century later, were not sufficiently developed to detect the parallax of any of the stars.

It was not until 1838 that a German astronomer, Friedrich Wilhelm Bessel (1784-1846), succeeded in measuring the tiny parallax of a star called 61 Cygni, which is actually a pair of stars revolving about one another. The

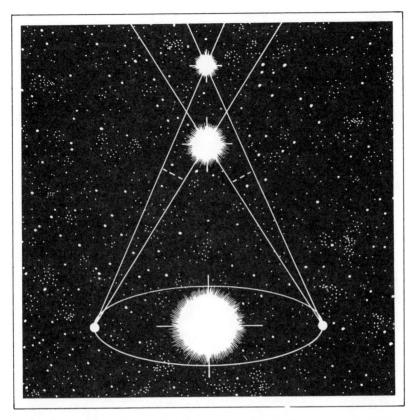

From one side of the orbit, the nearer star is slightly to one side of the farther one; from the other side of the orbit, the nearer star is slightly to the other side.

two, even observed together, are not particularly bright, but the star has an unusually high proper motion, which is why Bessel chose it to study. It turned out to be some 106 trillion kilometers (64 trillion miles) away from Earth. Because light travels 9.46 trillion kilometers (5.88 trillion miles) in one year, that distance is one "light-year." The star 61 Cygni is thus 11.2 light-years away from us.

At about the same time that Bessel was accomplishing this feat, a Scottish astronomer, Thomas Henderson (1798–1844), measured the distance of Alpha Centauri and found it to be about 4.3 light-years away. Alpha Centauri—two stars revolving about each other, with a third star a long way off—is the closest known star to Earth.

Astronomers are, more and more, using the *parsec* as a unit of distance, and that is equal to 3.26 light-years or 31 trillion kilometers (19.2 trillion miles). Alpha Centauri is, then, about 1.3 parsecs from us, and 61 Cygni about 3.4 parsecs.

The stars, in other words, were seen to be just what Nicholas of Cusa had thought them to be, four centuries before. If not infinitely many, they clearly existed in vast numbers. And the stars were suns, all at immense distances, strewn widely across vast stretches of space.

Man's perception of the heavens had altered irrevocably. Virtually nothing remained of ancient astronomy.

Modern Novas

In 1838 in South Africa, the English astronomer John Herschel (1792–1871) studied the stars near the south celestial pole, stars that were never visible from European latitudes. Herschel noted a bright first-magnitude star in the constellation Carina, a star known as "Eta Carina." Earlier astronomers who had traveled to the southern hemisphere had noted it as only a weak fourth-magnitude star.

Could it be a nova? Apparently. As the years passed, it faded slowly, but then in 1843 it flared up again and for a brief period reached a magnitude of −1, becoming almost as bright as Sirius itself. Then it gradually faded to the sixth magnitude. It was not and is not quite a nova but rather a very irregular variable of an unusual type that we will return to later.

The first undoubted nova detected after the invention of the telescope was observed in 1848 in the constellation Ophiuchus by the English astronomer John Russell Hind (1823-1895). It occupied the same constellation as Kepler's nova, but its location was clearly different so that it was not a renewed flare-up of the earlier one. What's more, the new nova (the first since Kepler's) was unspectacular. Even at its brightest, it failed to attain the fourth magnitude.

Three or four more unremarkable novas were observed during the remainder of the 1800s. One of them, in the constellation of Auriga (and therefore called "Nova Aurigae"), was detected in 1891 by a Scottish clergyman, T. D. Anderson.

He was an amateur astronomer, making one among many interesting astronomical discoveries that have been made by amateurs. Anderson spotted Nova Aurigae despite its weak fifth magnitude star shine. To recognize a new star that low in radiance, Anderson must have memorized the exact configuration of almost every visible star in the sky.

With the dawn of the twentieth century, it had been almost three hundred years since any nova had been seen that was as bright as the first magnitude, if we except the dubious case of Eta Carinae.

But on the night of February 21, 1901, T. D. Anderson discovered his *second* nova while walking home. This one shone in the constellation of Perseus, and was therefore called "Nova Persei." Anderson notified the Greenwich Observatory of his discovery at once, and the professional star watchers turned their telescopes upon it immediately. Anderson had caught it early, for a wonder, and it was still brightening. Two days later, Nova Persei reached a peak magnitude of 0.2, as bright as Vega.

39

By that time, astronomers had arrived in the era of photography, which gave them an enormous practical advantage over their predecessors. Had the area of the sky within which Nova Persei shone been photographed before its appearance?

Indeed it had. At Harvard Observatory, that same sky region had been photographed only two short days before Anderson made his new discovery. In the very place where Nova Persei was now shining, the Harvard photographs showed a very dim star of the thirteenth magnitude, only about 1/630 of the brightness necessary to make it just visible to a sharp-sighted but unaided eye.

In four days, Nova Persei had brightened thirteen magnitudes and had increased its brilliance some 160,000 times. Almost at once, it started to fade in an irregular fashion and, after several months was again invisible to the unaided eye. Eventually, it returned to the thirteenth magnitude.

Some seven months after Nova Persei had blazed forth, photography proved to be useful in another way. To the eye alone, even at the telescope, the star seemed just a star. However, if a photographic film, instead of the eye, were placed at the focus of the telescope and a long exposure were made, enough light would accumulate to reveal a dim fog of light around Nova Persei that gradually, over the weeks and months, grew in size. The expansion was that of the light the star emitted during its brilliant phase, moving outward in all directions at the speed of light and illuminating the cloud of thin dust and gas about the star. By 1916, fifteen years later, it was possible to observe a dim ring of thicker gas about the star, the gas itself seeming to have been ejected at the time of the brightening of the star and now to be expanding in all directions, though at speeds considerably less than that of light.

It seemed clear that the star had suffered a titanic explosion that had expelled gases as well as having produced a flash of light. That much was evident, though astronomers did not then know anything about the events that transpired inside the star or what mechanism might cause a stellar explosion. But they could give the phenomenon a name—Nova Persei was an example of an "eruptive variable" or an "explosive variable." Perhaps all novas were eruptive variables and such an expressive and accurate name ought to be substituted for "nova." It would have been useless to try; the expression "nova" has stayed fixed to the mind from the day Tycho first used it, and there is every indication that it will remain.

A still brighter nova was seen by several different observers on June 8, 1918, in the constellation Aquila. At that time, it was a first-magnitude star, and two days later it was at its peak, shining with a magnitude of −1.1, or with almost the brightness of Sirius.

Nova Aquilae appeared during World War I, and in earlier centuries many would have considered it an omen. In fact, it was so considered by some even in the twentieth century. The war was nearing its end, and in the spring of 1918 the Germans launched a great offensive in France, a final gamble to win. They staked all their last reserves on it, and made fearsome gains—but not quite enough. By early June, the Germans had run out of steam, and the French and British were now being rapidly reinforced by increasing numbers of American troops. It was clear that Germany was through, and, indeed, within five months she surrendered. Nova Aquilae was called "the star of victory" by the Allied soldiers at the front.

Again, Harvard Observatory photographs showed the star, before its outburst, to have been a dim star that varied somewhat between the tenth and eleventh magnitudes. In five days, it had brightened 50,000 times, but, as was to be expected, it faded quickly. By September it could barely be seen by the unaided eye. After eight months, it could only be seen by telescope.

Nova Aquilae is the brightest nova to appear in the sky since 1604, and nothing as bright has appeared since. Brightness, however, is not the only way in which a nova can make its mark.

There was a growing feeling that novas arose from faint undistinguished stars. If one simply looked at a star that later became a nova, the pre-novas seemed to have nothing about them that was in any way remarkable. On the other hand, one could do more than look at a star.

Astronomers, by the end of the 1800s, had the spectroscope, which was capable of spreading out the waves of light in the order of the length of those waves. This produced a rainbow of light: red, orange, yellow, green, blue, and violet (in order of decreasing wavelength). From the distribution of the light; from the nature of any missing wavelengths, which showed up as "dark lines" crossing the spectrum; and from the position of those dark lines, it was possible for astronomers to tell whether a star was moving toward us or away from us, whether it was hot or cool, what its chemical constitution was, and so on.

In that case, what about the spectrum of a pre-nova?

Unfortunately, it is extremely difficult to get the spectrum of a faint star, and there are very many faint stars. It would be an enormous task, even with computers, to get the spectrum of all the stars in the sky, and indeed the spectra of only a small minority of them exist. Once astronomers grew interested in Nova Aquilae, they found that the original star from which it brightened did have a spectrum on record. To this day, Nova Aquilae is the *only* nova of any kind that has a spectrum on record taken before the star had begun to brighten.

The spectrum, however, showed nothing unusual about Nova Aquilae in its pre-nova stage, except that it seemed to be a hot star, with a surface temperature of about 12,000° C., as compared to 6,000° for our Sun. This made sense, for even without knowing the details of what went on inside a star and how it managed to explode in the course of nova formation, astronomers would expect that a hot star would be more likely to explode than a cool star.

In December 1934, a nova appeared in the constellation Hercules, and it is referred to as "Nova Herculis." Nova Herculis was, to begin with, a slightly variable star, hovering between the twelfth and fifteenth magnitude. Photographs, studied later, showed that on December 12, when it was brightening, it was still too dim to be seen by the unaided eye. On December 13, however, it was at the third magnitude and was seen by an amateur English astronomer.

It brightened rather slowly for a nova, but by December 22 it reached its peak magnitude of 1.4. It then declined irregularly, sinking slightly, then recovering in part, and by April 1 was just barely visible to the unaided eye. It then sank rapidly, and by May 1 it was down to the thirteenth magnitude, about where it had been at the beginning.

Astronomers might have felt justified, then, in turning to other star studies, but almost at once Nova Herculis began to brighten again. By June 2 it shone at the ninth magnitude. It continued to brighten rather slowly until September, when it was at 6.7 or nearly bright enough to be noticed by the unaided eye. It then decreased again very slowly, and it was not untill 1949, fifteen years after its first appearance, that it returned for the second time to the thirteenth magnitude.

It is increasingly clear that a nova need not be

43

viewed as increasing in brightness only once. In fact, there are known to be "recurrent novas." A nova in the constellation Corona Borealis peaked at the second magnitude in 1866, and then did precisely the same thing in 1946. There are novas that have reached peaks on three and even four occasions. It may be that Eta Carina is a recurrent nova, though it has even more interesting possibilities, as we shall see.

The most recent bright nova appeared in the constellation Cygnus on August 29, 1975. "Nova Cygni" brightened with unusual suddenness to the second magnitude through a range of perhaps as much as nineteen magnitudes. It brightened thirty-million-fold in a single day, but it then faded off rapidly and was lost to sight within three weeks. Apparently, the faster and more extreme the brightening, the faster and more extreme the dimming—though the tapering off is always slower than the earlier brightening.

How Luminous? How Common?

How much light do novas really emit? We talk about novas which approach this magnitude or that, which are as bright as Sirius or brighter than Venus, but that doesn't tell us everything. One nova can appear brighter than another because it is actually brighter (more "luminous") or because it is closer to us and therefore appears brighter than it really is.

One way or another, it is possible nowadays to estimate the distance of stars. Considering the brightness of a star at its actual distance, it is not difficult to calculate what the brightness would be were it at some other distance. It would seem dimmer as its distance increased and brighter as its distance diminished according to a simple rule: Brightness varies inversely as the square of the distance.

Thus, our Sun is by far the brightest star in the sky. Its magnitude is −26.91 as compared with −1.42 for Sirius, the next brightest star. The Sun is 25.49 magnitudes brighter than Sirius, where each magnitude represents a 2.512-fold increase in brightness. Therefore, the Sun shines in our sky with a brightness fifteen billion times that of Sirius.

However, our Sun is also incomparably the nearest star in the sky. It is only 150 million kilometers (ninety-three million miles) away from us, which is equal to a distance of 0.000005 parsecs. Sirius, on the other hand, is 2.65 parsecs away and is, therefore, 530,000 times as far away from us as the Sun is.

Suppose, now, that we viewed the Sun and Sirius from the same distance. (The standard distance used by astronomers for the purpose is ten parsecs.)

If we imagined the Sun to be ten parsecs away, it would be two million times as far away as it is. Its brightness would diminish by the inverse square law, 2,000,000 × 2,000,000 or 4,000,000,000,000 times. If we decreased the Sun's magnitude by dividing its brightness 2.512 times for each magnitude decrease, we would find that the magnitude of the Sun, allowing for a decrease of four trillion times, would be 4.69. At a distance of ten parsecs, then, the Sun would have a magnitude of 4.69. This is its "absolute magnitude." It would be a fifth-magnitude star and a fairly modest member of the heavenly community.

As for Sirius, which is 2.65 parsecs away, its distance from us would be increased only 3¾ times if we imagined it moved outward to a distance of ten parsecs. Its brightness would be decreased, but not by much, and at ten parsecs Sirius would have an absolute magnitude of 1.3. It would still be a first-magnitude star at that distance, though not among the brightest.

When we speak of "brightness," we speak of the magnitude a star actually has in the sky. If we want to compare the appearance of two stars as it would be if they were at the same distance from us—if we compare their absolute magnitudes, in other words—we would speak of "luminosity."

A comparison of the brightness of two objects depends in part upon distance, so that a burning match we are holding in our hands is brighter than Sirius. A comparison of the luminosity of two objects is the real thing; it tells us which object is really delivering more light and by how much.

At equal distances, Sirius is 3.4 magnitudes brighter than our Sun. That means it is twenty-three times as luminous as the Sun.

Where do the novas stand on this scale? It is not always easy to judge the distance of a particular nova since they are often quite far away; but from the information that has been obtained for a number of them, the average absolute magnitude before they brighten as novas is about 3, so that they tend, in the beginning, to be about five times as luminous as our Sun. At the peak of their brightness, the average absolute magnitude is −8, so that at its brightest a nova would become about 150,000 times as luminous as our Sun. This is only an average, of course.

Some astronomers distinguish between two kinds of novas: fast and slow.

Fast novas increase in luminosity 100,000 times or more in only a few days. The peak luminosity is maintained for less than a week, and then there is a steady, moderately rapid decline.

A slow nova increases in luminosity more slowly and

erratically, and by a lesser amount as well. It then declines even more slowly and erratically than a fast nova.

Nova Persei and Nova Cygni are examples of fast novas; Nova Aurigae and Nova Herculis are slow ones. Recurrent novas, or at least those that recur every few decades, tend to show smaller increases in luminosity than do ordinary novas—even including slow ones.

How common are novas?

Prior to 1900, they were hardly ever seen, but now they are seen more frequently. This is not because novas have increased in number, but simply because more people are watching and astronomers now have better techniques for detecting them. Even so, those we see are by no means all there are.

To understand why, let's begin by asking how many stars there are. With the unaided eye, we can see about 6,000, but with the telescope we can see many millions more.

Is the number infinite, as Nicholas of Cusa thought?

What argues against an infinite number is the Milky Way, our galaxy, an immense band of starlight that encircles our sky and that, through a telescope, proves to be a vast aggregation of very dim stars.

The total mass of the galaxy is about 100 billion times that of the Sun. However, most of the individual stars in the galaxy are considerably smaller and less massive than the Sun. Therefore, it may be that there are as many as 250 billion individual stars in the galaxy.

Astronomers estimate that there are about twenty-five novas a year in our entire galaxy, on the average. If this is compared to the total number of stars in the galaxy, we see that among all those stars only one in ten billion turns nova in any one year.

The fact that there may be twenty-five novas per year in the galaxy does not mean that we will see that

We are part of a huge pinwheel of stars, but with our unaided eyes we see only a few in our immediate neighborhood.

many, no matter how hard we look. The dust clouds that hide the center from us make it certain that we will see no novas that flare up near the center (where most of the stars are) or anywhere in the far half of the galaxy.

For that reason, we are only likely, at best, to see two or three novas per year by the light they emit.

3

LARGE AND SMALL STARS

Solar Energy

If we think of a nova increasing its luminosity 100,000-fold in a few days, we must be aware that it is pouring energy into space at an enormous rate. An average nova at its peak will emit as much energy in a day as the Sun will in half a year.

Where does that energy come from?

For that matter, before we can answer that question, we ought to ask where the Sun itself obtains its energy. The Sun has been shining for 4.6 billion years at more or less its present rate. In that time, it has expended an incredible total of energy, and yet it is still shining and will continue doing so in its present fashion for five or six billion years. Where does all that energy come from?

Until the middle of the 1800s, this question did not bother anyone a great deal. The people of ancient and medieval times thought the Sun was made of a special heavenly material that simply had the ability to shine. It couldn't stop shining any more than Earthly objects could stop deteriorating with time. Then, too, the Sun was not known to be so very old. It was thought to have been shining for only a few thousand years.

As the 1800s wore on, however, scientists grew a little uneasy over the matter. They didn't believe that heavenly bodies were fundamentally different from Earth in chemical constitution. They began to understand that the Sun was not thousands but millions of years old, and they were beginning to study the properties of energy more and more thoroughly.

In 1847, the German physicist Hermann L. F. von Helmholtz (1821–1894) worked out the "law of conservation of energy" as a consequence of the carefully studied behavior of those processes that involved energy changes. The law stated that energy could not be created or destroyed but could only change its form. Other scientists had gotten much the same idea in the 1840s, but Helmholtz produced the most persuasive arguments, and he usually gets the credit for the law.

What's more, Helmholtz was the first to turn his full attention to the problem of solar energy. The Sun could not get its energy from nowhere; it could not create the energy out of nothing. Where, then, did the energy come from?

Helmholtz tried various energy sources that were well understood. Could the Sun get it by ordinary chemical burning? Could it get the energy from the continual infall of meteoric material? His first attempts either supplied insufficient quantities of energy or else involved changes in the mass of the Sun that would have produced easily measurable results but did not.

Finally, in 1854, Helmholtz decided that the only known source of energy that could power the Sun and would involve no ruinous complications was the energy that came from its own contraction. Its own massive material was slowly falling inward, and the energy of that fall was converted into radiation and would energize the Sun for many millennia.

This wasn't altogether satisfactory, for if the Sun has been contracting for a few tens of millions of years, it must have started with a size so large that it would have stretched out to Earth's orbit. Earth could only have been formed when the Sun was substantially smaller than that, so Earth could only be some tens of millions of years old.

Toward the end of the 1800s, geologists and biologists both strongly suspected that the Earth, and therefore the Sun, was much older than a few tens of millions of years. The Earth had to be hundreds of millions of years old, at least, and perhaps a billion years or more. The Sun had to be at least that old, too, and solar contraction wouldn't supply nearly enough energy for that. What would, then?

As the century came to its end, a new source of energy was quite unexpectedly borne in upon the consciousness of humanity. In 1896, the French physicist Antoine-Henri Becquerel (1852-1908) discovered "radioactivity." He found that the atoms of the metal uranium broke down very slowly, but quite steadily, into other, smaller atoms.

In 1901, another French physicist, Pierre Curie (1859-1906), showed that radioactivity involved the production of small quantities of heat—very small quantities. Still, since radioactive changes could continue over billions of years, and considering the amount of radioactive substances in the Earth as a whole, the total quantity of heat produced would be enormous. It was clear that a new and very intense source of energy had been discovered.

In 1906, the New Zealand-born physicist Ernest Rutherford (1871-1937) showed that the atom was not just a tiny sphere, as had been thought, but was composed of still smaller "subatomic particles," chiefly (as we now know) *protons, neutrons,* and *electrons.* The protons and neutrons, which were relatively massive for such tiny particles, were located in the equally tiny nucleus at the very center of the atom. Around the atom there circled the light electrons. It was the nucleus that underwent changes and released energy during radioactivity, and so people began to speak, eventually, of "nuclear energy."

Well, then, could it be that the Sun shone because of nuclear energy?

The one source of nuclear energy known in the early decades of the 1900s was the radioactive breakdown of atoms of such substances as uranium and thorium. Could the Sun be a huge ball of uranium and thorium?

No, it could not be. By the early 1900s, the chemical constitution of the Sun was known through the use of the spectroscope, which was mentioned earlier in the book. Let's consider it again.

The Sun's light, when passed through a glass prism, is spread into a rainbow or *spectrum,* something that was first shown in 1666 by the English scientist Isaac Newton (1642-1727). This happened because light is made up of tiny waves of different lengths, and on passing through a glass prism each ray of light is bent to an extent that depends upon its particular "wavelength." The shorter the wave, the more it is bent. The spectrum consists, therefore, of all the light waves spread out in order, from the longest waves at one end to the shortest at the other.

In 1814, the German optician Joseph Fraunhofer (1787-1826) showed that the solar spectrum was crossed

52

White light from the Sun becomes a rainbow and modern optics is born.

by numerous dark lines. These dark lines (we now know) existed because the Sun's atmosphere absorbed some of the light of particular wavelengths that passed through it. The Sun's light reaches Earth with those wavelengths missing, therefore, and the gaps are the dark lines in the spectrum.

The German physicist Gustav Robert Kirchhoff (1824–1887) showed, in 1859, that every different kind of atom absorbed particular wavelengths of light that no other kind absorbed (or emitted those wavelengths when hot). By studying wavelengths that were absorbed or emitted, the identity of the atom that was absorbing or emitting the light could be determined.

In 1861, the Swedish physicist Anders Jonas Ångström (1814-1874) identified some of the dark lines in the solar spectrum with hydrogen, which is made up of the simplest atoms that exist. For the first time, a clear identification of at least part of the heavenly body was made—and it was constituted of a material that existed on Earth. So much, by the way, for Aristotle's notion that heavenly bodies were made up of substances unique to themselves.

Since then, the solar spectrum has been studied in greater and greater detail, and other types of atoms have been discovered in the Sun, all of which also exist on Earth. Even the proportions of the different atoms present can be determined. It is therefore possible to state quite assuredly that the Sun is *not* a ball of uranium and thorium. Those substances are present only in tiny traces and can produce only amounts of energy that are thoroughly insignificant in comparison with the amount the Sun radiates constantly.

Does that mean that nuclear energy cannot be the source of the Sun's energy?

Not at all. In 1915, an American chemist, William Draper Harkins (1873-1951), presented theoretical considerations for supposing that many types of nuclear rearrangements, other than ordinary radioactivity, could yield energy. He pointed out, in fact, that a nuclear rearrangement that yielded unusually high quantities of energy was one in which four hydrogen nuclei were converted into one helium nucleus. He suggested that such "hydrogen fusion," as it is now called, was the source of the Sun's energy.

The difficulty here was that radioactivity progressed

spontaneously on Earth and would surely progress just as spontaneously on the Sun, so that uranium breakdown could be a plausible source of solar energy only if enough uranium were present there. Hydrogen fusion, on the other hand, doesn't proceed under ordinary conditions but requires enormous temperatures, temperatures that even the glowing surface of the Sun does not begin to supply.

In the 1920s, however, Eddington considered the question of why the Sun did not collapse and contract into a small object under the pull of its own enormous gravity. The one force that could keep it expanded against gravity was heat, and he calculated how hot the solar interior would have to be in order to maintain the Sun at its actual size. It was clear that the temperature would have to be in the millions of degrees, and the figure currently accepted as the temperature at the center of the Sun is about 15,000,000° C.

What's more, in 1929 the American astronomer Henry Norris Russell (1877-1957) worked out the constitution of the Sun in greater detail than anyone had managed to do before. His analysis of the solar spectrum showed that about 75 percent of the mass of the Sun was hydrogen and the remaining 25 percent helium. These were the two simplest atoms. All the more complex atoms existed in the Sun to a total of 1 percent at most.

If the Sun is essentially a ball of hydrogen and helium, hydrogen fusion is the only possible nuclear reaction that could supply the necessary energy for solar radiation. Further, the solar interior, if not its surface, supplies a high enough temperature for the purpose.

In 1938, the German-American physicist Hans Albrecht Bethe (1906-   ) took into account the Sun's composition and its central temperature and worked out a sensible mechanism for what went on at the center of the

Sun. This has been refined since, but what happens, as nearly as can be determined, is that solar energy is derived from the fusion of four hydrogen nuclei into a helium nucleus, as Harkins had suggested a quarter of a century earlier.

What works for the Sun undoubtedly works for other stars, so that, having solved the problem of solar energy, we have in all likelihood solved the problem of stellar energy generally.

The process of hydrogen fusion can continue under equilibrium conditions, producing an unchanging (or a very slowly changing) energy output for different periods of time, depending on the mass of the star.

The more massive a star, the more hydrogen it will contain, but the more heat will be required to keep it expanded under the greater gravitational pull of that more massive star. Indeed, the need outpaces the supply as the mass increases. This means that the large fuel supply of a massive star is used up more quickly than the small fuel supply of a less massive one. The greater the mass of a star, the shorter its lifetime as a hydrogen-fusing device.

The fuel supply of a massive star is consumed so quickly that it can remain a normal star only for a few million years. A much smaller star will use its lesser hydrogen supply in so prudent a fashion that it can energize for some 200 billion years.

The Sun, which is intermediate in this respect, has a hydrogen supply that can endure for ten to twelve billion years. It has already existed 4.6 billion years, so it has not yet quite reached the midpoint of its life expectancy as a normal star.

Stars that are in this stage of their life cycle are said to be on the "main sequence." The Sun is on the main sequence. So are nearly 85 percent of the stars we see in the sky.

White Dwarfs

Not all stars are on the main sequence. The manner in which this was discovered begins in a way that seems to have nothing to do with the matter, and yet ends there, and goes on to explain the nature of novas. Here is how it happened.

It had always been assumed that stars were single objects. To be sure, there were a few places where some stars seemed closely grouped, but then people or trees might be closely grouped and still be independent, single objects.

Once the telescope was invented, it was possible to see that stars were sometimes grouped more closely than had earlier been imagined. Indeed, there was sometimes a pair of stars so near each other that the pair looked like a single star to the unaided eye. I mentioned earlier, for instance, that 61 Cygni and Alpha Centauri are "single stars" that turned out to be a pair of stars very closely combined.

Once it became clear that stars were distributed throughout vast stretches of space, however, it could be argued that, of two closely spaced stars, one might be fairly close to us and the other vastly farther away. The two stars might not be close to each other at all but would *seem* close because they were nearly in the same direction.

If the stars were distributed through space in a random way, there would be a certain chance that some would be more or less directly behind others and would look close together to us. In 1767, an English geologist, John Michell (1724-1793), argued that the number of very close stars was considerably higher than might be expected of a random distribution. He therefore suggested that stars actually existed in pairs.

Goodricke, in 1782, may have been encouraged by Michell's argument to suggest that Algol was actually a pair of stars, circling each other, so that one periodically eclipsed the other—but that was only a reasonable suggestion rather than an actual observation.

William Herschel (who later worked on the general shape of the galaxy) was, in the 1780s, studying stars that were very close to each other. He hoped that one would be nearby and one far away so that he could determine the parallax of the near one compared to the far one and get the distance of the near one.

Instead of finding a parallax, however, he discovered that, in many cases, the two stars were clearly moving about each other. He could actually observe them doing so. Ordinary double stars might be so only in appearance, but here Herschel had *binary* stars (from a Latin word for "pair") that were *really* close to one another, so close that they were held in each other's gravitational field. Each circled the center of gravity of the pair.

At first, binary stars were thought to be rare, but the more astronomers studied the stars, the more binaries they found. It is now thought that up to 70 percent of the stars that exist are part of binary systems or systems that are still more complicated. It is single stars like our Sun that appear to be in the minority.

The discovery of one particular binary star led eventually to an important advance.

Bessel, who had been the first to determine the distance of a star, was studying the changing position of Sirius in order to measure *its* distance. He noted that Sirius' changing position was not of the type one would expect of a parallax. Instead, it progressed in one direction in a wavy line. The waviness made it appear that the gravitational attraction of some nearby object was forcing Sirius into an elliptical orbit. This orbit, com-

58

bined with its straight-line proper motion, produced the waves.

For a star like Sirius to be forced to travel in noticeable waves meant that the gravitational pull of the other object had to be enormous. The other object had to be a star; nothing less would do. Bessel could not see anything where that star should be, however, and in 1844 he came to the conclusion that Sirius was a binary star with a "dark companion." The companion, he felt, was a star that was invisible because it had burned itself out, so that it moved through space as a blackened cinder of what it had once been.

In 1862, an American telescope maker, Alvan Graham Clark (1832–1897), was preparing a new telescope and was testing it on Sirius to make sure that it gave a sharp image. It did, but near Sirius was a speck of light. Clark, thinking this meant there was a defect in the instrument, tested the lens carefully and found it to be perfect.

Studying the speck of light, Clark found it to be in the position where Bessel's "dark companion" would have to be if it were to be responsible for Sirius' wavy motion. The obvious conclusion was that it *was* the companion.

The companion has a magnitude of 8.4, so it is not "dark," but there wasn't much of a change in calling it Sirius' "dim companion." Nowadays, we call Sirius "Sirius A" and its dark or dim companion "Sirius B."

In 1893, the German physicist Wilhelm Wien (1864–1928) showed that it was possible to determine the surface temperature of a star from the details of its spectrum. In 1915, the American astronomer Walter Sydney Adams (1876–1956) managed to study the faint spectrum of Sirius B and found that its surface temperature was surprisingly high. Sirius B was hotter than our Sun, though not quite as hot as Sirius A.

If Sirius B were hot—and it had a surface temperature of 10,000° C.—each bit of its surface must glow brilliantly, more brilliantly than an equal-sized portion of the Sun's surface. Why, then, was Sirius B so dim? It could only be that its surface was extremely small in extent. The star was glowing brilliantly, but there was little of it to glow so its total light was small.

Nowadays, it is believed that Sirius B is only 11,100 kilometers (6,900 miles) in diameter. This makes it a little smaller than Earth, with its diameter of 12,756 kilometers (7,950 miles).

However, it is small only in size. Bessel knew it was there without actually seeing it because of its gravitational effect on giant Sirius A. That gravitational effect had not changed simply because astronomers had discovered that Sirius B was only as large as a small planet. From its gravitational pull, they have calculated that it has about 1.05 the mass of our Sun—all that mass squeezed into its small, less-than-Earth size.

The average density of the Earth (if we imagined the whole planet stirred into a homogeneous mass) is about 5,500 kilograms per cubic meter. Sirius B, however, has a density 530,000 times as great.

The average density of Sirius B is, therefore, about 3 billion kilograms per cubic meter. An American twenty-five-cent coin, made up of matter like that in Sirius B, would weigh about 1,900 kilograms (4,200 pounds).

Sirius B does not, however, have the same density all the way through. It is least dense near its surface and grows increasingly dense with depth so that it is most dense at its core. (This is true of any astronomical body, including Earth and Sun.) The density of Sirius B at its center may be as high as 33 billion kilograms per cubic meter.

When it was first discovered that Sirius B was very

small, it was at once obvious that its density was much greater than that of the densest object on Earth. This would, some years earlier, have appeared to be ludicrous, but by the time Adams had made the key discovery about Sirius B's temperature, it was already understood that an atom consisted of an extremely dense and tiny nucleus surrounded by nearly massless electrons. It was therefore suggested by Eddington in 1924 that in an object such as Sirius B, the atoms were smashed and the nuclei were driven far closer together than they could be in matter composed of intact atoms.

Matter consisting, in this fashion, of smashed atoms and nuclei driven closely together is called "degenerate" matter. The Sun has temperatures and pressures so high in its deep interior that its center contains degenerate matter. A star like Sirius B, however, is composed almost entirely of degenerate matter.

The surface gravity of any object depends upon the mass of that object and the distance from its surface to its center (that is, its "radius"). For instance, the mass of the Sun is 333,500 times that of the Earth. The Sun's radius, however, is 109.1 times the Earth's radius so that at the Sun's surface, one would be 109.1 times as far from the center of the body as one would be at the Earth's surface. The greater distance from the center tends to weaken the pull of gravity that one would experience on the Sun's surface.

To determine the Sun's surface gravity, its mass must be divided by the square of its radius. That is $333,500/(109.1)^2$, which comes to about 28. In other words, the surface gravity of the Sun is about twenty-eight times that of the Earth.

If we think of Sirius B, however, we must remember that though the mass is 1.05 times that of the Sun, the radius of the little star is much smaller than that of the

Sun. The distance from the surface of Sirius B to its center is only 0.008 times that of the Sun's radius. The surface gravity on Sirius B, then, is $[1.05/0.008)^2] \times 28$, or 470,000 times that of the Earth.

Because Sirius B is white-hot in temperature and yet is so small, it is an example of a *white dwarf.* Because it is a star of so high a density and small a size, it is an example of a *collapsed star.*

Sirius B and all white dwarfs are stars that are no longer on the main sequence. On the main sequence, fusion reactions in the center develop heat that keeps the

The star the size of a planet—a white dwarf.

star an extended body. Once the fusion reactions fail, the star can't remain extended, and its own gravitational field forces it to collapse into a white dwarf.

Up to about 15 percent of the stars in the galaxy may be white dwarfs. That means there are perhaps as many as forty-five billion white dwarfs in the galaxy. Because of their small size, they are so dim that only those that are comparatively near to us can be seen. Even Sirius B, which is the white dwarf that is closest to us, would not be visible without a telescope even if the blinding light of nearby Sirius A were not present.

Red Giants

The white dwarfs seem now to be an essential key to the puzzle of nova formation—but not by themselves. There is another type of star we must deal with, a type that is also not on the main sequence.

When the Danish astronomer Ejnar Hertzsprung (1873-1967) first worked out the main sequence in 1905, he noticed that there were two kinds of red stars. Some were very dim and some were very bright; there were *no* red stars of intermediate brightness.

A red star is red because its surface is cool, or at least no more than red-hot, whereas stars like our Sun are white-hot. Red stars may have surface temperatures of only about 2,000°C. One would expect such stars to deliver comparatively little light per unit surface, and, if they were only the size of the Sun or smaller, they would have to be dim. Dim red stars are, therefore, no surprise. How do we explain the very bright red stars, however?

In order for a cool star to be very bright, it must

63

make up for the fact that not much light is emitted per unit surface by having a very large surface—*much* larger than that of our Sun. Bright red stars must have diameters perhaps a hundred times that of our Sun. Such stars, like Betelgeuse or Antares, are therefore called *red giants.*

When the main sequence was first worked out, it was clear that red giants were stars that were not on it. It seemed reasonable to suppose that red giants were stars that were in the process of birth, that they were slowly condensing under their own gravitational field and that

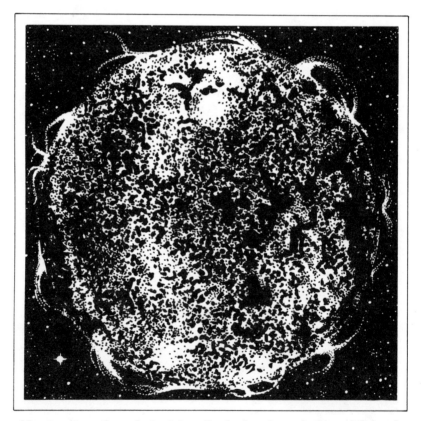

Monster star—the red giant (our Sun's size shown in lower left-hand corner).

in so doing they were growing smaller and hotter. Eventually, the red giants would condense to normal size and temperature and would then enter the main sequence.

That thought is no longer accepted. Scientists have studied clusters of stars in which all were thought to be the same age, since the entire cluster had very likely been formed simultaneously. Astronomers realized that all the stars of the cluster were evolving and that the more massive the star, the more rapid the evolution. They, therefore, determined the mass of various stars and had, as it were, a series of "stills" showing different stages of evolution. The most massive stars were red giants, which showed that, though such a star was indeed not on the main sequence, it was a late stage in stellar evolution and not an early one.

How is it that a red giant forms?

The answer is currently thought to be this: Slowly, over the millions and billions of years, the hydrogen at the core is consumed, and the helium that is formed by fusion, being denser than hydrogen, collects at the very center of the Sun. Hydrogen fusion continues at the outer edge of the growing ball of helium at the center, but it is on the helium that we must focus our attention.

As the helium at the center condenses under its own weight, the helium ball becomes steadily smaller, denser, and hotter. It eventually develops high enough temperatures and pressures to allow "helium fusion" to begin. Helium nuclei combine with each other to form the more complex nuclei of carbon, nitrogen, and oxygen.

In the process, heat is delivered to the star, over and above that being produced by the usual process of hydrogen fusion at the outer edge of the helium core. This causes the outer layers of the star to overheat and to expand enormously, much more so than in a normal star subsisting entirely on hydrogen fusion. The expanding star can be considered to be leaving the main sequence at this point.

As the outer layers expand, they cool to a mere red heat, but the expansion of the surface more than makes up for that. If the star's diameter increases by 100 times, its surface area increases 100 × 100 or 10,000 times, so that the total heat it radiates is far higher than that of most normal stars, despite its cool surface.

Helium fusion delivers far less energy than hydrogen fusion does, so that the helium supply runs short in much less time than the hydrogen supply might have. The products of helium fusion can go on to fuse further, but all the energy available from helium fusion is perhaps only one-twentieth that of hydrogen fusion—and the red giant continues to emit energy at a fearsome rate.

This means that the red giant stage does not endure very long from a stellar viewpoint. From a human viewpoint it does, for it may last one or two million years. And that is why there are relatively few red giants to be seen. Only about 1 percent of the stars in the galaxy are red giants. That means only about 2.5 billion in the entire galaxy, and, of course, we can only see the portion in our part of the galaxy, even though they could be seen for long distances if it were not for the dust clouds. Most stars either haven't reached the red giant stage or have passed beyond it.

The nuclei at the center of a red giant continue to fuse until the temperature can no longer rise high enough to allow for new fusions. In the most massive stars, the temperature can rise very high indeed, but, even so, fusion can only proceed until iron nuclei are formed. The iron nuclei signal a dead end. Whether iron nuclei are broken into smaller ones ("fission") or whether they undergo fusion into larger ones, no energy is produced. Indeed, in either case, energy must be *supplied*. We might consider the iron nuclei to be the final "ash" of the fusion reactions that go on in a stellar interior.

Whether the core of the red giant reaches a temperature beyond which its mass can no longer drive it, or whether it finally produces iron nuclei, the end is the same. The nuclear fire goes out and there is nothing to keep the star expanded against its own gravity—and it collapses. It does so very rapidly, too.

As the star collapses, it heats up, and some of the hydrogen that still remains in the outer region may undergo sufficient heating and compression to allow it to fuse. There is thus an explosion that tends to hurl some of the star's substance into space, so that an expanding sphere of gas and dust can surround the collapsing star.

Some stars that are visible to us are in this stage. The expanding sphere of gas is lit up by the star and we see it most clearly at the edges, where the greatest thickness is in our line of sight. The collapsed star looks as though it were surrounded by a smoke ring.

Any cloud of dust and gas in interstellar space is called a *nebula* (the Latin word for "cloud"), and when a nebula seems to be a ring around the star, resembling a planet's orbit, it is called a *planetary nebula*.

About 1,000 planetary nebulas* are known, of which the most famous is the Ring nebula in the constellation of Lyra.

At the center of each planetary nebula is a very hot, blue-white star (as one would expect a freshly formed white dwarf to be), whose radiation continues to drive the shell of gas outward. The shell grows ever larger, thinner, and fainter, until it disappears into the widely scattered gas and dust of interstellar space. What is left,

---

* As in the case of "nova," the proper plural of "nebula" is "nebulae," but these days the Anglicized "nebulas" is more and more commonly used.

then, after a period of 100,000 years or so, is a white dwarf with no detectable nebula about itself—the stage in which Sirius B now is.

A white dwarf no longer has nuclear fusion going on within it, so it has no further source of heat. Very slowly, over the ages, the white dwarf cools down. Finally, it radiates too little light to be seen, and it becomes a *black dwarf*. The universe may not be old enough even yet for many black dwarfs, or any at all, to have formed.

Binaries and Collapse

Does it seem, now, that we can guess what happens when a star goes nova?

When a red giant collapses, there is a flash of light as the hydrogen in its outer layers condenses. Should not that flash of light represent the nova? Then, too, the explosion ejects gas and dust and could we not see such an ejection of gas and dust in Nova Persei and Nova Aquilae?

Actually, this is not so. Studies of pre-novas (the few that have been made) show that they were not red giants. What's more, after a nova has dimmed and returned to its original state ("post-nova"), it is not a white dwarf. Both before and after, the star appears to be a main sequence star, somewhat brighter and hotter than the Sun.

To work out this puzzle, let's remember that most stars are members of binary systems. Since this is so, we are entitled to wonder what happens when one of the members of a binary comes to the end of its stay on the main sequence, expands to a red giant, and then collapses to a white dwarf, while the other member of the binary remains on the main sequence.

Both members of a binary star must have formed at the same time, and the more massive of the two should finish its stay on the main sequence sooner and therefore be the one of the pair that first ends as a white dwarf.

Yet the white dwarf we know best, Sirius B, seems to defy this conclusion. It is no longer on the main sequence, though it is only 1.05 times the mass of the Sun, while Sirius A, which is still on the main sequence, is 2.5 times the mass of the Sun. How do we explain that anomaly?

The reasonable conclusion is that Sirius B *was* the more massive star to begin with and moved into the red-giant stage first for that reason. When the Sirius B red giant collapsed, a large portion of its mass was ejected. The result was that the portion of Sirius B that finally condensed into a white dwarf was considerably less massive than the original star had been.

What's more, much of the ejected matter at the time of the collapse of Sirius B may have been trapped by Sirius A, which meant that the latter became more massive than it had been originally. (It also means that Sirius A's lifetime as a main-sequence star was thereby seriously shortened.)

There's nothing there that seems to indicate that a nova had ever formed in Sirian binary, but the notion of mass transfer from one member of a binary to the other turned out to be important.

The key discovery concerning novas that led to the present understanding of the phenomenon came in 1954.

By that time, post-novas were being carefully studied, and one finding was that many of them seemed to flicker. They showed rapid tiny variations of light, not at all like the steady output of normal stars. Naturally, astronomers were looking for something, *anything* that would mark post-novas off from ordinary stars, and this flicker seemed hopeful.

69

One of the stars under observation was Nova Herculis, or, rather, the star that had been Nova Herculis twenty years before and was thereafter named DQ Herculis. In 1954, the American astronomer Merle F. Walker noted that superimposed on its flickering was a definite dimming that lasted for an hour, followed by a brightening to the original level. Further observation showed that this dimming took place periodically, every 4 hours and 39 minutes.

It became clear that DQ Herculis was an eclipsing binary, as Algol was, which was something no one had expected. The reason it hadn't been noticed before was that the change was not great, and the period was so short that no one was prepared for so rapid a repetition of the change—and so no one was watching for it. In fact, when DQ Herculis was recognized as a binary star, it had the shortest period associated with one up to that time.

This meant that the two stars of the binary revolved about a common center of gravity unusually quickly, and that meant, in turn, that they were extremely close. In fact, the best estimate, these days, is that the two stars of DQ Herculis are separated by not much more than 1.5 million kilometers (900,000 miles), center to center. If both stars were the size of our Sun, they would be nearly touching.

Was this just a coincidence? Could the fact that DQ Herculis was a very close binary have nothing to do with the fact that it had recently been a nova?

The thing to do was to study other post-novas to see if they, too, were very close binaries. Out of ten post-novas studied by Walker's colleague, Robert P. Kraft, seven showed definite signs of being very close binaries.

Naturally, it would be too much to expect of coincidence that all the binary systems would be seen edge on, so that one would get in the way of the other and produce an eclipse. The post-novas that did not show any signs of an eclipse were nevertheless seen to be close binaries from a careful study of the spectral lines.

Ultra-close binaries are very rare, and novas are also very rare. That so many stars are both novas *and* ultra-close binaries simply cannot be blamed on coincidence. There has to be a connection.

Then another fact was discovered. The post-novas seemed to be ordinary main-sequence stars, but close study of the spectra showed the presence in addition of small, white-hot stars that had to be white dwarfs. In other words, the post-novas seem all to be ultra-close binaries in which one of the stars is a white dwarf.

That had to be why the change in brightness was so small in the course of the eclipse. When the white dwarf gets in front of its normal companion, it hides virtually none of it, so there is little decrease in total brightness as compared with the situation where both stars are shining unobscured. When the companion gets in front of the white dwarf, it obscures a star whose *total* brightness, however white-hot it might be, is not great. Again, there is little decrease in total brightness.

From this combination of a white dwarf and a main sequence star in an ultra-close binary, astronomers worked out what must happen to produce a nova.

To start with, the ultra-close binary is made up of two main-sequence stars. The more massive of the two (A) finally becomes a red giant. As the red giant expands, it gets so large as to touch its companion (B), which captures some of the outer layers of A, becoming more massive and, therefore, shorter lived. Eventually, A collapses to a white dwarf, while B continues its now-shortened stay on the main sequence.

Before long (as stellar lifetimes go), B begins to run low on fusion fuel and starts an expansion. Before this

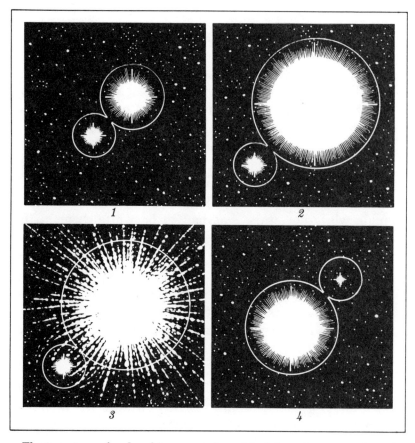

The two stars of a close binary are bound to interact, one growing at the expense of the other. Here is what may have happened in the case of the Sirius binary.

can go to extremes and before B can become an outright and recognizable red giant, its outer layer gets close enough to white dwarf A for some of B's matter to spill over into A's region of gravitational influence.

When it happened earlier, in reverse, A's matter collided with B's surface because both were normal stars. Now, however, B's matter doesn't collide with A's surface because A is a white dwarf and very small. Instead, B's matter goes into orbit about white dwarf A and forms an *accretion disk*.

It is so named because the matter in orbit interferes with itself, thanks to mutual collisions of particles and atoms, so that as a result of a sort of internal friction, portions of it lose energy and sink toward the white dwarf. These portions slowly spiral inward, and the white dwarf gradually gains masses of material (a process called *accretion*) on its surface.

Although the hydrogen at the core of B is gone, and it is expanding toward the red giant stage, B's outer layers, which leak across, are still almost entirely hydrogen. The white dwarf A, which has very little hydrogen of its own, even in its outer layers, is thus steadily collecting hydrogen from its companion.

The hydrogen that reaches the surface of the white dwarf is compressed by the intense surface gravity of that tiny star and is heated up in consequence. More and more hydrogen arrives, and it continues to heat up. Eventually, the temperature reaches the point where some of the hydrogen undergoes fusion, and the surface of the white dwarf heats up further.

Eventually a further point is reached where, between the heating up of the hydrogen and of the surface of the white dwarf, things become sufficiently hot to initiate a vast fusion reaction in the accretion disk. A large portion of the disk fuses, producing an enormous flash of light and other radiation and driving the upper layers of the accretion disk beyond the gravitational influence of the white dwarf.

The enormous flash of light is what we see from Earth as a nova, and the portion of the accretion disk that is driven away is the cloud of dust and gas that we can see expanding about the post-nova.

The fusion process gradually dies down, the activity ceases, and, over a long period of time, the white dwarf's surface cools. The cycle then starts again as hydrogen from B continues to leak over, rebuilding the accretion disk which slowly approaches the cooling surface of A. Eventually there is another explosion. The nova can, in this way, recur many times before B completes its expansion and is ready to condense into a white dwarf. (Binaries are known in which both members are white dwarfs—although, if they are sufficiently far apart, neither one may have had a history as a nova since there wouldn't be enough leakage from one to another.)

Generally, the first nova explosion is the brightest, and the result is sometimes called a *virgin nova*. Nova Persei, Nova Aquilae, and Nova Cygni may have been virgin novas. The second explosion may not come for as long as 20,000 years, and it is less bright. Further recurrences are steadily less spectacular.

The white dwarf itself contributes to the intensity of the nova reaction. The white dwarf has massive nuclei—such as those of carbon, nitrogen, and oxygen atoms—at its surface, and small amounts of these can mix with the incoming hydrogen. Massive nuclei tend to hasten the hydrogen fusion. If more than an average amount of these massive nuclei mix with the hydrogen, the fusion speeds through the hydrogen shell particularly rapidly, producing a much brighter initial flash and a consequent more rapid fade-off. If the carbon, nitrogen, and oxygen are added in comparatively small quantities, the fusion ignition is relatively slow so that the initial flash is not as bright and fades off more slowly. That is why there are fast novas and slow novas.

The requirement for nova formation is thus rather stringent, and it is clear that very few stars in the galaxy would qualify. What is required is a binary star, and an ultra-close one at that.

Most particularly, then, our Sun does not qualify. It is not part of an ultra-close binary. It is not, in fact, part of a binary of any kind, as far as we know. Eventually, after five billion years or more, the Sun will have consumed enough of its hydrogen and will begin helium fusion. At that time, it will begin to expand to a red giant and, eventually, collapse to a white dwarf—but it will do so in single blessedness and without interference. It will never become a nova.

4

BIGGER EXPLOSIONS

Beyond the Galaxy?

Not all novas are ultra-close binaries involving a white dwarf. Perhaps one in a thousand is not, but it involves a totally different type of phenomenon. To understand such exceptions, we will have to broaden our view of the universe.

When it first became clear that the stars we see in the sky are part of a structure that possessed a fixed shape of finite size—the galaxy—it was taken almost for granted by most astronomers that this structure included all, or almost all, of the stars there were. The galaxy, in other words, just about made up the universe.

It was thought that the only objects in the sky that

could possibly be considered to lie outside the galaxy were the "Magellanic Clouds." These are located deep in the southern sky and are not visible from European latitudes.

The first Europeans to see and describe them (in 1520) were part of the expedition to the Far East by Ferdinand Magellan, who traveled by means of a western route. To reach the Far East, the expedition (which eventually completed the first circumnavigation of the globe) had to get past the Americas, and to do that they had to sail far south and pass through what is now known as the Strait of Magellan. From those far southern latitudes, the Magellanic Clouds are seen high in the sky.

The Magellanic Clouds are two regions of dim light that look as if they might be small, detached portions of the Milky Way. Because they are detached, it might also be that they are not part of the galaxy, of which the Milky Way is, so to speak, the rim.

As time went on, the Magellanic Clouds were shown to consist of large numbers of very dim stars, just as the Milky Way was. By the 1930s, it became clear that the Large Magellanic Cloud was 47,500 parsecs away, while the Small Magellanic Cloud was 50,500 parsecs away. Both were well beyond the confines of the galaxy.

Also, both were far smaller than the galaxy. Whereas the galaxy is made up of about 250 billion stars, the Large Magellanic Cloud may only have as many as ten billion and the Small Magellanic Cloud no more than two billion.

The Magellanic Clouds could be considered as small satellite galaxies of our galaxy, which can now be distinguished from other structures of the sort as the "Milky Way" galaxy. One might argue that the Magellanic Clouds had somehow become detached and that the two, with the Milky Way galaxy, made up one gravitationally bound system—just as one might speak of the Earth-Moon system as a unit.

The question then arose: Is there anything that exists outside the Milky Way-Magellanic system?

Through the 1800s, few astronomers thought anything did. In fact, there was only one object that looked as if it might and did not appear to be a star.

After all, not everything in the sky is a star or a dimly luminous object, like the Milky Way or the Magellanic Clouds, which can be resolved into stars. Some astronomical objects are, so to speak, of a different species altogether.

Thus, in 1694, the Dutch astronomer Christian Huygens (1629-1695) described a bright, fuzzy object that, to the unaided eye, was seen as the middle star of the three that, to imaginative folk, made up the sword of the giant hunter who was pictured in the sky as the constellation of Orion. The telescope showed it as a region of luminous fog surrounding half-obscured stars.

This was almost at once assumed to be exactly what it appeared to be. It was a nebula, a vast cloud of dust and gas, illuminated by the stars that shone within it. It is called the "Orion nebula," and we now know it to be about nine parsecs across and some 500 parsecs away from us. It is a thin and rarefied cloud by Earthly standards, a better vacuum than we can make in the laboratory, but the widely-spread particles mount up in the line of sight and suffice to obscure the stars within the nebula.

There are other bright nebulas to be seen, many quite beautiful in form and color. Nor are they found in the galaxy only. In the Large Magellanic Cloud, there is the "Tarantula nebula," which is far larger than the Orion nebula.

There are also dark nebulas. William Herschel, in studying the Milky Way closely, noted that there were regions in which few or no stars were to be seen. He took these at face value and assumed that they represented regions that did not contain stars and that the Earth happened to be so situated that human beings could see into the empty regions as though looking into a tunnel. He described such regions as "holes in the heavens."

By 1919, there were 182 such dark regions recorded, and it soon seemed unlikely there could be so many holes in the crowded galaxy, all pointing toward Earth. The American astronomer Edward Emerson Barnard (1857-1923) and the German astronomer Maximilian F. J. C. Wolf (1863-1932) suggested independently, in the 1890s, that these regions were nebulas that, unlike the Orion nebula and others of the sort, did not shine because they did not happen to enclose stars that would serve to illuminate the dust particles.

Such dark nebulas were visible only because they were located in the same line of sight as thick fields of stars. The nebulas obscured the stars and showed up as black, irregular shadows against them.

Dark nebulas that did not contain stars, and luminous nebulas that did, were not all the nebulas that could be seen in the sky. There were some that did not fall into either class and represented potential puzzles. The most prominent and brightest of these, and the only one visible to the unaided eye, has the appearance of a dim, somewhat fuzzy "star" of the fourth magnitude. It is in the constellation of Andromeda and was noted by some of the Arab astronomers.

In 1611, it was first looked at through a telescope by the German astronomer Simon Marius (1573-1624), and he is usually listed as the discoverer of what came to be called the "Andromeda nebula."

A French astronomer, Charles Messier (1730-1817), was an avid hunter of comets, which were temporary phenomena that appeared, changed position against the starry background, and finally disappeared. In 1781, he

79

made a catalog of fuzzy objects in the sky that were *not* comets but were permanent inhabitants of the sky and that maintained a fixed position among the stars. His intention was to make sure that other comet seekers would not mistake them for comets and be disappointed. The Andromeda nebula was thirty-first on the list, and it is sometimes known as "M31" in consequence.

The Andromeda nebula was puzzling because it was not a dark nebula, but was luminous. On the other hand, there was no clear reason for the luminosity since no stars seemed to be present within it to serve as a reason for its shining. A cloud of dust and gas that shone without stars seemed an anomaly.

The Messier list contained other examples of patches of luminous fog without stars. Some of these were eventually resolved into stars by astronomers such as Herschel, who showed that some Messier objects were dense spherical clusters of stars (*globular clusters*). A few could not be so resolved, however.

Presumably, if an explanation could be found for the Andromeda nebula, that same explanation would hold for other smaller, less prominent nebulas as well. What, then, was the Andromeda nebula?

Two totally different kinds of explanation were offered before the 1700s were over.

First, the reason that no stars were visible in the Andromeda nebula might be that, like the Milky Way or the Magellanic Clouds, the nebula consisted entirely of stars rather than dust, but with stars that were too dim to be seen.

If so, the postulated stars of the Andromeda nebula must be exceedingly dim, for although telescopes managed to resolve the fog of the Milky Way and the Magellanic Clouds into swarms of faint stars, they did not do the same for the Andromeda nebula. In even the best telescopes of the day, the Andromeda nebula remained a fog.

The most reasonable way of explanation was to suppose that the Andromeda nebula was *so* far away that even telescopes fell short of showing the stars that made it up, since those stars would appear far dimmer than those making up nearer objects such as the Milky Way and the Magellanic Clouds. And if the Andromeda nebula was excessively far away, and yet visible even to the unaided eye, it must be an enormous cloud of stars indeed.

This was the point of view of the German philosopher Immanuel Kant (1724-1804). In 1755, he spoke of the existence of "island universes." Eventually, when the existence of the galaxy was recognized, it made sense to suppose that Kant's island universes could only be other, very distant galaxies, if they existed.

Kant's view was ahead of its time. Astronomers would not be ready to expand their vision beyond the galaxy and imagine the existence of numerous other galaxies for another century and a half. Less visionary, and therefore more acceptable, was the second view, that of the French astronomer Pierre-Simon de Laplace (1749-1827). He suggested, in 1798, that the solar system was, to begin with, a vast, spinning cloud of gas and dust that slowly condensed and, in the process, cast off lesser rings of dust and gas out of which the planets were formed. As the cloud condensed, the central regions eventually grew hot enough to shine and illumine the whole of the regions of dust and gas out of which the planets were forming. As the outer regions of the cloud became the planets, in other words, the central region became the Sun.

Kant had made a similar suggestion in the same book in which he spoke of island universes. Laplace went into greater detail, however, and pointed out that the Andromeda nebula might be viewed as an example of a planetary system in the process of formation. By this view, the Andromeda nebula was indeed a fog of gas and dust, but at its center was a star that was just beginning to shine and was not yet visible as such but that served to illuminate the whole.

Because of its use of the Andromeda nebula as an example, Laplace's notion was called the "nebular hypothesis."

If Laplace were correct, then the Andromeda nebula, as a single planetary system, must be reasonably close to appear as large as it does and would certainly be a part of the galaxy.

Throughout the 1800s, Laplace's view was the one generally accepted. Very few astronomers, if any, ranged on the side of Kant.

During the 1800s, the Andromeda nebula continued to grow less unique. As the skies were searched with better and better telescopes, it turned out that there were quite a number of nebulas that were luminous and yet showed no signs of stars.

The Irish astronomer William Parsons, third Earl of Rosse (1800-1867), paid particular attention to these nebulas and built what was then the world's largest telescope to help him in his studies. The telescope was almost useless because the weather on his estate was so bad he could scarcely ever get any viewing done. Now and then, though, he had a chance to study the nebulas, and he noted, in 1845, that a number of them seemed to have distinctly spiral structures, almost as though they were tiny whirlpools of light set against the black background of space.

The most spectacular example of this was M51, the fifty-first item on Messier's list. It looked like a pinwheel, and it soon came to be known as the "Whirlpool nebula."

Astronomers began to speak of "spiral nebulas" as a not uncommon class of objects in the sky.

Other nebulas were elliptical in outline, with no signs of spiral arms, and were called "elliptical nebulas." Both spiral and elliptical nebulas were markedly different from nebulas such as the one in Orion, which were filamentous and irregular in shape.

By the second half of the 1800s, it was becoming possible to take photographs of objects in the sky, even of dim ones. The camera had to be fixed to a telescope that was arranged to move automatically with the sky in order to neutralize the Earth's rotation on its axis. In this way, a long-exposure photograph could be taken.

In the 1880s, a Welsh amateur astronomer, Isaac Roberts (1829-1904), took a large number of photographs of nebulas. This was important because the camera could see and record, more objectively, the fine structure of these objects. Astronomers would no longer have to rely exclusively on the sometimes dubious artistic ability of observers trying to draw what they saw.

In 1888, Roberts was able to show that the Andromeda nebula had a spiral structure. This had not been noted before because the Andromeda nebula was seen much more nearly edge-on than the Whirlpool nebula. The spiral formation so evident in the latter case tended to be obscured in the former.

Roberts pointed out that, if nebulas were photographed periodically over a period of years, small changes relative to surrounding stars might show that the nebula was rotating at some measurable speed. That alone would show, unequivocally, that the nebula was a relatively small object and therefore relatively nearby. Anything as far away as one of Kant's island universes would have to be so huge it would take millions of years for it to make one rotation, and no measurable change in appearance could be detected in any reasonable time of study. In 1899, Roberts claimed that his photographs *did* show such rotational changes in the Andromeda nebula—and that seemed to be that.

Also in 1899, the spectrum of the Andromeda nebula was taken for the first time. It was found to be very similar to those of stars generally, whereas irregular clouds of dust and gas, like the Orion nebula, produced spectra that were completely different from those of stars and that usually consisted of a number of separated bright lines of distinct color. This meant that the Orion nebula and others of its type often showed delicate colors, whereas the Andromeda nebula and others of its type were white—and were, therefore, sometimes called *white nebulas*.

The spectrum of the Andromeda nebula made sense if Laplace were right and if the nebula were a developing star. In 1909, in fact, the English astronomer William Huggins (1824-1910) announced that his studies showed the Andromeda nebula to be a planetary system in a late stage of development.

There simply seemed no room for disagreement.

And yet, a difficulty that had come up toward the end of the century stubbornly refused to go away. It involved novas.

S Andromedae

On August 20, 1885, the German astronomer Ernst Hartwig (1851-1923) noted a star in the central regions of the Andromeda nebula. It was the first ever seen in connection with the nebula.

It is possible that some astronomers may have originally thought that the developing planetary system, which the Andromeda nebula supposedly represented, had finally reached its climax. The central region was no longer merely glowing but had finally caught fire and turned itself into a full-fledged sun. Had this been so, the star should have remained glowing and become a permanent fixture in the sky—but it did not.

Slowly the star faded away, and finally it disappeared in March 1886. It was a nova, clearly and without mistake—Nova Andromedae. It has since come to be referred to as "S Andromedae," and I will use that name for it.

But what was a nova doing in the Andromeda nebula? Could a single, developing star go nova before it had become a true star? And, if it had, then when the nova faded away, how was it that the Andromeda nebula remained as before, without a single visible change?

Yet who was to say that the nova was actually part of the Andromeda nebula? It might simply have been observed in the same line of sight as the nebula which, in actual fact, was shining well beyond it and was in no way affected by it.

Part of the nebula or not, however, S Andromedae was a very feeble excuse for a nova. Even though, at that time, astronomers had seen very few novas, they had seen enough to know that S Andromedae was unusually dim. Even at its brightest it reached a magnitude of only 7.2, so that it was never visible to the unaided eye. No one could possibly have stepped out of doors, seen S Andromedae in the sky, and remained standing, transfixed, thinking "Incredible! A new star!" as Tycho must have done three centuries earlier.

No one but a few astronomers at their instruments saw S Andromedae. Even they would very likely not have noticed it but for the accident that it shone in the featureless fog at the center of the Andromeda nebula, where no distinguishable star, however faint, had ever been seen before.

Photographs were taken of the Andromeda nebula that revealed the nova shining within it, but no spectra were taken of it. Spectra of dim objects were not easy to take in those days. The rapid brightening and slow dimming of S Andromedae was typical of a nova, however, and the only question one might ask about it was why it was so faint.

That question might not seem a very compelling one. Novas, after all, might simply come in widely varying brightnesses. At its peak, it might be extremely bright, like Tycho's, or rather undistinguished, such as the nova Hind had detected in 1848, which was only of the fourth magnitude at best. Nova Andromedae was merely less distinguished, that's all.

Since there was nothing then known of the causes and nature of novas, it was possible to argue that it might all depend on how luminous a star was to begin with. A particularly luminous star would blaze out unbelievably; a less luminous star would be more modest in its glow; an extremely dim star might not become visible to the unaided eye even at the peak of its brightness as a nova.

And so S Andromedae was dismissed. It had appeared and disappeared, had been noted and was forgotten.

Until 1901, that is. In that year, Nova Persei appeared and had briefly shone as a star of 0 magnitude. From the manner in which light seemed to be expanding within the ring of dust about it, astronomers could calculate the distance to Nova Persei. After all, they could see the apparent speed of the light, and they knew the real speed, so it was not difficult to determine the distance at which light would appear to move in the observed fashion.

They concluded that Nova Persei was at a distance of thirty parsecs from Earth.

This is not really far for a star. There might be a few thousand stars closer, but there are many billions further. The thought had to arise that the only reason that Nova Persei was as bright as it was was due to its proximity.

Could it be that all novas reached more or less the same level of luminosity—the same absolute magnitude—but that they were different in apparent brightness only because of distance?

For instance, suppose S Andromedae reached only a magnitude of 7.2 because it was farther off than Nova Persei. If the two novas had the same peak absolute magnitude, then S Andromedae would have to be about 500 parsecs away to shine as dimly at its peak as it did.

If so, the Andromeda nebula would have to be 500 parsecs away, if S Andromedae were part of it. If S Andromedae were merely in front of the nebula, then the Andromeda nebula would have to be more than 500 parsecs away, and perhaps considerably more than 500 parsecs away.

Even if the Andromeda nebula were no more than 500 parsecs away, it could not be a single planetary system in the process of formation. No single planetary system could be 500 parsecs away and look as big in the sky as the nebula did.

Astronomers refused to accept this reasoning, which was based, after all, on the supposition that Nova Persei and S Andromedae had the same peak luminosity. It seemed easier to suppose that they were of different peak luminosities and that S Andromedae did not merely *appear* to be very dim compared to Nova Persei but *was* very dim. S Andromedae could then be quite close as astronomical distances go, much closer than 500 parsecs, and so could the Andromeda nebula.

And, in that case, the Andromeda nebula could still be a planetary system in the process of development.

The Andromeda Galaxy

One American astronomer, Heber Doust Curtis (1872-1942), did not accept this easy way out. Suppose that S Andromedae were far away and that the Andromeda nebula was farther off than was supposed—much farther off, even. Might not the Andromeda nebula be so far off, then, that Kant's idea of a century and a half earlier was correct, that the Andromeda nebula was an island universe—an independent galaxy of stars well outside our own?

If so, the Andromeda nebula ought to consist of very, very many, very, very dim stars. Among them, novas ought to flare up now and then. While the stars in their courses would not be made out within the Andromeda nebula by even the best telescopes then available, any that brightened to a nova might conceivably become telescopically visible, even easily visible, as S Andromedae was.

Beginning in 1917, Curtis *did* discover novas in the Andromeda nebula, dozens of them. That they were novas was unquestionable, for they appeared, then faded away; then new ones appeared and faded away.

There were two important things about this crowd of novas. The first was that it *was* a crowd. In no other part of the sky did so many novas appear in one discrete area.

This meant they couldn't simply be appearing in that direction of the sky, independent of a nebula that would just happen to be lying indifferently behind them. If that were the case, why should such numbers be appearing *only* in that direction? It was too much to ask of coincidence that a unique gathering of novas and the Andromeda nebula should both just happen to lie in the same direction and to have no palpable connection. Curtis felt quite safe in supposing the novas to be lying within the nebula.

Why so many novas? Well, if the Andromeda nebula were an island universe and an independent galaxy, it might well have as many stars within it as our own galaxy has. There should therefore be as many novas occurring within its boundaries, even though it seemed but a small patch of light to our eyes, as in our own galaxy, which filled all the rest of the sky.

In fact, *more* would be seen in the Andromeda nebula than in our own galaxy. Curtis noticed that in the Andromeda nebula there were patches of darkness around the edges which, if the Andromeda nebula were actually a galaxy, might be large stretches of dark nebulas, clouds of gas and dust that obscured the stars behind it.

And the same phenomenon might also occur in our own galaxy. In addition to the small dark patches in the Milky Way, there might be much larger ones we knew nothing about (and, in time, this was found to be so), so that much of the thick star-regions of our own Milky Way might be totally invisible to us. Among those vast hidden crowds of stars (far greater in number than those we could see) would be many novas each year hidden by veils of cloud dust. Where the Andromeda nebula was concerned, however, we could see past the dust clouds from our lateral vantage point so that few of its novas would be hidden.

And, in fact, more novas were seen in the Andromeda nebula than in the rest of the sky.

The second interesting fact about the Andromeda novas was their extreme faintness. They could just barely be seen, even at their brightest and even with a powerful telescope.

If they were anything like ordinary novas, such as Nova Persei, they had to appear exceptionally dim because they were extraordinarily far away and that, too, fit in with the concept of the Andromeda nebula as an independent galaxy.

Curtis was convinced, and he became the outstanding astronomical spokesman for the idea of island universes.

He did not, however, have it all his own way. The concept remained difficult to accept, especially since there seemed to be new evidence that the Andromeda nebula was actually a nearby object. The Dutch-American astronomer Adriaan van Maanen (1884-1946) had interested himself, in particular, in measuring tiny motions of astronomical objects, including those of a number of spiral nebulas. Van Maanen corroborated Roberts' earlier observation that the Andromeda nebula had a measurable rate of rotation. He reported, in fact, that not only it but several other spiral nebulas had measurable rates of rotation.

We now know that van Maanen's measurements were incorrect for some reason. He was measuring changes in position that were just barely within the ability of his instruments to detect, and either something went slightly wrong with those instruments or his firm belief that there *should* be detectable rotation influenced his observations.

Nevertheless, van Maanen had an excellent reputation and, on the whole, deserved it, so that people tended to believe him. And, if the Andromeda nebula showed visible motion, it *had* to be close by regardless of dubious reports of crowds of all-but-invisible novas.

One of those involved in the controversy was the American astronomer Harlow Shapley (1885-1972). Shapley had recently made use of Cepheid variables as a way of measuring distances, a technique developed in 1912 by the American astronomer Henrietta Swan Leavitt (1868-1921). In this way, Shapley was able to show that the true center of our galaxy was far from the solar system and that we on Earth were in the outskirts of the galaxy. Shapley was the first to determine what we now believe to be the true size of the galaxy, instead of falling short, as all previous estimates had done. In fact, Shapley's estimate was originally somewhat too high. He was also the first to determine the distance of the Magellanic Clouds.

It might seem that Shapley, who had stretched out the distances within the galaxy and immediately outside it to new and unprecedented lengths, would be ready to imagine still other objects that were even further away. He was, however, a close friend of van Maanen, and he accepted the latter's results. He became the leading spokesman for the small-universe concept. The galaxy and the Magellanic Clouds were, in his view, all there were, and the various white nebulas were merely part of these structures.

On April 26, 1920, Curtis and Shapley held a well-publicized debate on the matter before a crowded hall at the National Academy of Sciences. Undoubtedly, Shapley had the greater reputation and represented the majority view, but Curtis was an unexpectedly effective speaker and his novas, in their dimness and their number, were a surprisingly powerful argument.

Objectively, the debate ended up to have been considered a stand-off, but the fact that Curtis had won even so much as a draw was an astonishing moral victory. As a result, there developed a steadily growing opinion (especially in the later light of hindsight) that he had won the debate.

The debate did not, in fact, decide the issue, though it seemed to have converted a number of astronomers to the island-universe viewpoint. What was needed was additional evidence, one way or the other—evidence that was stronger than anything that had yet been advanced.

The man who supplied it was the American astronomer Edwin Powell Hubble (1889-1953), who had at his disposal a new and giant telescope with a mirror 100 inches wide—the most far-seeing anywhere in the world up to that time. It was put into use in 1919, and, in 1922, Hubble began to use it to make time-exposure photographs of the Andromeda nebula and similar space objects.

On October 5, 1923, he found, on one of his photographs, a star in the outskirts of the Andromeda nebula. It was not a nova. He followed it from day to day, and it turned out to be a Cepheid variable. By the end of 1924, Hubble had found thirty-six very faint variable stars in the nebula, twelve of them Cepheids. He also discovered sixty-three novas in the Andromeda nebula, much like those Curtis had earlier detected.

Could all these stars be independent of the Andromeda nebula and, somehow, happen to be lying in the same direction? No! Hubble reasoned, as Curtis had, that so many very faint Cepheid variables could not be strewn in the direction of the Andromeda nebula simply by coincidence. A similar number was not to be found in any other comparable sky region.

Hubble felt he had detected the stars that made up the Andromeda nebula where previous astronomers had failed. He had succeeded because he had a superior telescope that outstripped all those that had gone before it.

Nor could Hubble's view be denied. Once the Andromeda nebula had been resolved into stars (only the few brightest, but those few were enough), the former notion of the nebula as a nearby object and as a planetary system in the process of formation was dead forever.

What's more, once Hubble had discovered Cepheid stars in the Andromeda nebula, he could make use of the Leavitt-Shapley method to determine its distance. His figures showed the nebula to be 230,000 parsecs away, about five times the distance to the Magellanic Clouds. The Andromeda nebula was clearly far outside the galaxy. It was also, clearly, a galaxy in its own right.

For a while, the various white nebulas were called *extra-galactic nebulas*, but eventually the word "nebula" was dropped as entirely inappropriate. The objects came to be called simply *galaxies*, and the Andromeda nebula became the "Andromeda galaxy" and has so remained ever since. In the same way the Whirlpool nebula became the "Whirlpool galaxy," and so on.

To hammer the final nail into the coffin of the small-universe view, Hubble showed, in 1935, that van Maanen's observations on the measurable rates of rotation of various galaxies had been erroneous.

And the other white nebulas, smaller in appearance and dimmer than the Andromeda, are all galaxies also, and, in general, further than Andromeda, some much further. The universe is now perceived as a vast assemblage of galaxies, of which our own Milky Way galaxy is but one.

As a matter of fact, Hubble's estimate of the distance of the Andromeda galaxy (and, therefore, of all further galaxies) was low. In 1942, the German-American astronomer Walter Baade (1893-1960) demonstrated that there were two kinds of Cepheid variables that had to be used in different ways to establish cosmic distances. The correct kind had been used by Shapley in determining the size of our galaxy and the distance of the Magellanic Clouds.

Hubble had, however, unknowingly used the other kind in his estimate of the distance of the Andromeda galaxy, and his calculations were therefore mistaken. When corrected, it was realized that the Andromeda galaxy was 700,000 parsecs from us, fourteen times as far away as the Magellanic Clouds.

Supernovas

Every solution produces new puzzles. Once astronomers had agreed that the foggy patch in Andromeda was an exceedingly distant galaxy, it became necessary to take another look at S Andromedae, which had created so little stir back in 1885.

It had been argued that if S Andromedae were as luminous as Nova Persei, it would have to be about 500 parsecs away in order to be no brighter than seventh magnitude at its peak. But what if it were as far away as the Andromeda galaxy was now known to be?

If the Andromeda galaxy were at the distance of Hubble's first estimate of 230,000 parsecs, S Andromedae would have had to be about 200,000 times as luminous as Nova Persei to attain seventh magnitude brightness at such a distance. Since the Andromeda galaxy is actually at a distance of 700,000 parsecs, S Andromedae would have to have a luminosity about two million times that of Nova Persei at its peak, or about twenty billion times the luminosity of our Sun.

The Andromeda galaxy, we now know, has about twice the mass of ours, or about the mass of 200 billion stars like our Sun. It should, perhaps, have a total luminosity of about 100 billion stars like our Sun (assuming that most stars are considerably less luminous than the

94

Sun). If S Andromedae at its peak was twenty billion times as luminous as our Sun, it was one-fifth as luminous as the entire vast galaxy of which it was part.

If this were so, S Andromedae could not be looked upon as just another nova. It created a luminosity about a million times, and perhaps two million times, that of an ordinary nova.

Most astronomers, however, found this information hard to accept. Some die-hard opponents of the large-universe idea argued that the Andromeda galaxy *couldn't* be a distant galaxy, because if it were, S Andromedae would be impossibly luminous.

Others took the less combative position that the exceedingly faint novas detected by Curtis and Hubble were indeed novas of the Andromeda galaxy but S Andromedae was not. They said it was at a distance considerably less than a thousandth that of the Andromeda galaxy, the 500 parsecs that had once been calculated, and that was why it seemed so much brighter than the other Andromeda novas. It just happened to lie in the direction of the Andromeda galaxy. For *one* nova to become so bright was not asking too much of coincidence.

Hubble, however, totally disagreed. He held firmly to the belief that S Andromedae was part of the Andromeda galaxy and that it was an abnormally bright nova.

How could one decide?

The Swiss astronomer Fritz Zwicky (1898-1974) reasoned as follows. Suppose that S Andromedae was indeed abnormally luminous. Such a phenomenon would likely be very rare, for it is the common experience of humanity that phenomena that present an extreme of something fairly ordinary are rare roughly in the proportion that they are extreme. It would therefore represent a waste of time to track the Andromeda galaxy for another nova like S Andromedae. There were, however, so many galaxies known to exist that to have an abnormally luminous nova in some *one* of them would not be rare at all. What's more, since such an abnormally luminous nova was nearly as bright as the entire galaxy of which it was a part, there would be no problem seeing it. A nova of the S Andromedae type, in any galaxy however distant, could be seen if the galaxy itself could be seen.

As a matter of fact, since S Andromedae first appeared, about twenty-one novas had been detected in or near one or another of what were known to be galaxies. They had always been too dim to see with the unaided eye (as they would have to be if they were located in distant galaxies) and had not been much studied in consequence. To Zwicky, these seemed to have been what he would have been looking for.

In 1934, only fifty years ago as I write, Zwicky began a systematic search for what he now called *supernovas*, a term he was the first to use. He focused on a large cluster of galaxies in the constellation Virgo and, by 1938, had located no fewer than twelve supernovas in one or another of the galaxies of that cluster. Each one, at its peak, was almost as bright as the entire galaxy itself, and every one of them had to be shining with the luminosity of billions of times that of our Sun.

Could all twelve objects be deceptive? Could they all be relatively close novas that happened to be seen in the direction of one or another of the galaxies of the Virgo cluster? It was logically and mathematically unreasonable that so wild a coincidence could take place. Astronomers began to accept the fact that the novas were actually inside the galaxies that appeared to surround them and that they were supernovas.

Soon, additional supernovas were discovered in succeeding years by Zwicky and by others. By now, some 400 supernovas had been detected in various galaxies.

From the numbers that have been observed, it seems

reasonable to conclude that in any given galaxy, on the average, one supernova explodes every fifty years. One supernova irrupts, in other words, for every 1,250 ordinary novas.

It is now estimated that within 300 million parsecs, there are about 100 million galaxies that can be made out with our present telescopes and within which, therefore, a supernova can be seen when it appears. If each galaxy averages one supernova in every fifty years, that means that a supernova explosion appears in one or another of the visible galaxies every fifteen seconds!

Unfortunately, we can't see them all. Some would be obscured by vast dust clouds in their own galaxy or be eclipsed by banks of other, less luminous stars that lie between a supernova and ourselves. And, of course, there aren't enough astronomers to keep a constant eye on every one of the hundred million visible galaxies, in any event.

Nevertheless, 400 supernovas have been detected in other galaxies during the last fifty years. That represents one supernova every 6½ weeks, on the average.

Supernovas are clearly incredible objects of astoundingly explosive character. Were it possible for our Sun to go supernova, it would vaporize every planet in the solar system by the time it had reached peak brilliance.

Were it possible for Alpha Centauri, which is only 1.3 parsecs away, to go supernova, it would blaze in our day and night sky with a brightness that, at maximum, would be 15,500 times that of the full Moon, or about one-thirtieth that of the Sun.

So it is completely understandable that astronomers can think of few things they would like to study in full detail so much as a supernova, and it is frustrating indeed that the star watchers are forced to study them in other galaxies at distances of 700,000 parsecs or more.

While no sane person could wish a supernova to erupt

at too close a distance, it isn't unreasonable to hope that there might be one ready to ignite in our own Milky Way galaxy, at a distance of, say, 700 parsecs rather than 700,-000 or more.

And, with supernovas exploding in particular galaxies every fifty years or so, surely there must have been some in the Milky Way galaxy in the past.

Indeed there were! Looking back with the wisdom of hindsight, it seems clear that there have been at least four undoubted supernovas in our Milky Way galaxy during the past thousand years.

The first was the nova in Lupus in 1006, the one that was about a tenth as bright as the full Moon; it may have been the brightest nova to shine in the sky during man's time on Earth. Then there was the nova in Taurus in 1054; the nova of 1572 studied by Tycho; and the nova of 1604 studied by Kepler.

Only four? Considering the every-fifty-year average, there might well have been twenty.

The difficulty is, of course, that we can't yet see our entire galaxy, only the portion closest to ourselves. In the *visible* portion, we might average only one in 250 years. For instance, there is evidence, which we'll return to later, of a supernova that could have been visible in the sky in 1670, but which no one reported. Undoubtedly, it was masked by dust clouds.

There's another bad break. If only four supernovas in the Milky Way galaxy were visible in our skies during the last thousand years, why was the fourth and last in 1604? The telescope was invented five years later!

The closest supernova since 1604 is S Andromedae, 700,000 parsecs away. It was seen by telescope and it was photographed, but its spectrum was not studied. And, in a century, there has been nothing closer.

Too bad!

5

SMALLER DWARFS

The Crab Nebula

A supernova is so tremendous an explosion that it is hard to believe it doesn't leave any trace of itself behind. A star that shines briefly with all the light of an entire galaxy of stars must surely leave ashes—and it does.

Since the existence of supernovas has been known only since the 1930s, one can scarcely expect the ashes to have been recognized for what they were. Those ashes might, however, have been noticed earlier without their nature being recognized.

In 1731, for instance, an English astronomer, John Bevis (1693-1771) was the first to report a small, fuzzy patch in the constellation of Taurus.

Messier, the comet hunter, was aware of it, too; and fearing that it might carelessly be thought to be a comet, he placed it on his list of objects to be disregarded by other comet hunters. In fact, he placed it first on the list, so that the fuzzy patch of Taurus is sometimes known as "M1."

The first to examine M1 in detail was Lord Rosse, in 1844, working with the same large telescope he was soon to use to detect the spiral nature of many distant galaxies. To him, M1 was not just a mass of fuzz. His telescope showed it to him more clearly than that; it looked, rather, very much like a turbulent volume of gas, something that almost forced an interpretation of itself as a remnant of a violent explosion. Within the gas were numerous ragged filaments of light that looked, to Rosse, something like the legs of a crab. He named M1 the "Crab nebula," and that is what it has been called ever since.

The Crab nebula began to attract considerable attention because nothing else in the sky looked quite like it. Nothing else looked so clearly as if an explosion were in progress. It began to be photographed and, of course, that meant it became possible to compare photographs taken over the years.

The first to do so was an American astronomer, John Charles Duncan (1882-1967). In 1921, he took a photograph of the Crab nebula and compared it carefully with one taken in 1909 by another American astronomer, George Willis Ritchey (1864-1945), who had made use of the same telescope that Duncan was now using. It seemed to Duncan that the Crab nebula was slightly larger in his photograph than it was in Ritchey's. Apparently, it was expanding.

If that were so, it might well be that the nebula was the remnant of a nova and, from the quantity of dust and gas, a rather large nova. Another photograph that Duncan took in 1938 made the matter unmistakable.

Soon after the first report of expansion in 1921, Hubble (who was soon to resolve the Andromeda galaxy into stars), judging from this and from the location of the Crab nebula in Taurus near where the Chinese had reported a "guest star," suggested that the nebula was the still-expanding remnant of the bright nova of 1054.

It might be, but how to demonstrate the fact?

From the observed rate of expansion of the nebula, one could calculate backward to see when all the dust and gas had been together in a tiny point of light. That would tell astronomers how long a time had elapsed since a star at the site of the Crab nebula had exploded. The period of time since the explosion turned out to have been about 900 years.

That placed the explosion almost exactly in 1054, the year of the bright nova in Taurus. Since then, astronomers have universally accepted the identification of the Crab nebula and the nova of 1054.

It was possible to convert the apparent rate of expansion of the Crab nebula into a true rate by studying the displacement of the dark lines of its spectrum. The figure turns out to be about 1,300 kilometers (800 miles) per second. One can then easily calculate how far off the Crab nebula must be for that true rate to yield the apparent rate of expansion as measured in photographs. It turns out that the Crab nebula is some 2,000 parsecs from us.

Knowing the distance, one can then calculate from the apparent width of the Crab nebula that the cloud of dust and gas is now about four parsecs in diameter and is, of course, continuing to widen steadily.

From the report of how bright the nova of 1054 was and the knowledge of its actual distance, it is possible to calculate that, viewed at ten parsecs, the standard distance for determining absolute magnitude, the nova at its peak brightness would have been shining with an absolute magnitude of −18. At its peak, then, that starburst would have shone with something like 1.6 billion times the luminosity of our Sun, or about one-sixtieth as bright as the entire Milky Way galaxy, if that brightness could be concentrated into a point. The 1054 nova was, beyond dispute, a supernova.

Since the Crab nebula is 2,000 parsecs away, it must be a true nebula consisting of dust and gas. It could not be a very distant collection of stars as the Andromeda nebula turned out to be. In that case, the Crab nebula ought to emit a spectrum that consisted of separate bright lines of different wavelengths, as the Orion nebula does. This was not so, however. The Crab nebula has a continuous spectrum, emitting light at all wavelengths as stars do. In fact, it is at a considerably higher temperature than stars are, for the Crab nebula emits light at very short, energy-intense wavelengths, including not only ultraviolet light but also the shorter-wave x-rays and even the still shorter-wave gamma rays. It also produced copious quantities of long-wave radio radiation that is *polarized*, oscillating in one direction only.

The source of such a continuous and energetic spectrum was mystifying until 1953, when a Soviet astronomer, Iosif Samuilovich Shklovskii (1916–   ), suggested that it originated with high-speed electrons moving through a strong magnetic field. The result of such electron motion would be radiation of just the sort that was observed. This was not merely theory. Precisely this sort of phenomenon (on a tremendously smaller scale, of course) is observed in connection with certain particle accelerators called *synchrotrons* that are designed by nuclear physicists. In those, electrically charged particles are forced through magnetic fields and give off what is called *synchrotron radiation*.

It seemed, then, that the Crab nebula was producing

synchrotron radiation on a vast scale, but where did the electrons come from? Where did all the energy derive that drove electrons through the magnetic field, during all the nine centuries since the supernova exploded?

If 1945, Baade, who had worked out what is now believed to be the true distance of the Andromeda galaxy with the German-American astronomer Rudolph L. B. Minkowski (1895–1976), had observed small changes in the Crab nebula near two stars in the center of its structure. They maintained that one of those two stars must be a remnant of the original object that had undergone the supernova explosion. Still, to keep such a stream of synchrotron radiation going, that remnant star must be emitting energy at a rate 30,000 times that of our Sun. How this could happen was a puzzle that was not to be solved for another quarter of a century.

If the 1054 supernova left such an amazing remnant of itself, other supernovas might have done the same. Any expanding cloud of dust and gas exhibiting synchrotron radiation would be highly suspect. The difficulty is, though, that the longer ago a supernova took place, the wider and more rarefied the expanding cloud and the less intense the radiation.

It would appear that the reason for detecting the unusual properties of the Crab nebula is that the 1054 supernova was comparatively recent, reasonably close, and in clear view. There were no interposed dust clouds to speak of.

Still, radio waves can penetrate dust clouds without trouble, and, after World War II, astronomers had developed the instruments and the techniques to detect radio waves without difficulty and with steadily increasing delicacy.

In 1941, Baade detected nebulous filaments in the constellation of Ophiuchus at about the location where

Kepler had reported the supernova of 1604. That supernova remnant is not much more than one-third the age of
the Crab nebula, but it is also much further from us, some
11,000 parsecs away, so that it is that much harder to discern. Baade had no ready way of being certain that those
filaments of dust and gas were really supernova remnants. But, in 1952, two astronomers at Cambridge Uni-
versity, R. Hanbury Brown and Cyril Hazard, found them
to be a strong source of radio-wave radiation. That tied it
in quite clearly with the supernova of 1604.

In that same year, Brown and Hazard detected radio
waves from the region in Cassiopeia corresponding to
Tycho's nova. Later, Minkowski, using the 200-inch telescope on Mt. Palomar in California, found traces of a visible remnant at the sight. These traces are about 5,000
parsecs from us. Then, in 1965, a radio-wave source was
located in the constellation of Lupus that must be a remnant of the great supernova of 1006, which may have been
only 1,000 parsecs away.

Thus, the four known supernovas of the last thousand years all left behind remnants. In fact, there is a
fifth remnant. In 1948, two British astronomers, Martin
Ryle (1918-1984) and F. Graham Smith (1923-   ), detected an intense radio source in Cassiopeia. Later, Minkowski detected the nebulosity that went along with it,
called "Cassiopeia A." It was not at the site of Tycho's
supernova, but it seemed to have the properties that
matched a supernova remnant. If it had been caused by a
supernova, such an explosion should have been visible on
Earth about 1677, but it must have been obscured by interstellar clouds, for no one reported it.

Another suspicious entity, called the "Cygnus loop,"
is, as you might guess, in the constellation of Cygnus. It
consists of curved bits of nebulosity that seem to be part
of a ring that is 3° in diameter, or six times the width of

the full Moon. If it is the remnant of a supernova, that starburst must have exploded some 60,000 years ago.

Still another remarkable structure first came to astronomers' attention in 1939, when the Russian-American astronomer Otto Struve (1897–1963) detected a faint nebulosity in the southern constellation of Vela. From 1950 to 1952, the finding was followed up by the Australian astronomer Colin S. Gum (1924–1960), who published his results in 1955.

It turns out that the Gum nebula, as it is called, is the largest one known, taking up perhaps one-sixteenth of the entire sky. It is so rarefied, however, that it is not easily seen and is, in any case, too far South to be tracked well from Europe or the United States.

The Gum nebula is roughly spherical and is about 720 parsecs in diameter. Its center is about 460 parsecs from the solar system, which makes it the closest supernova remnant we know of. Its near edge is only 100 parsecs away, and, for a time, astronomers even suspected the solar system might actually be within the nebula.

It may be the result of a supernova that exploded 30,000 years ago and that may well have shone as brightly as the full Moon for a short while. Modern man was just coming into existence then. We can wonder whether they and Neanderthal men took awed notice of this second moon in the sky, assuming they were sufficiently far south to see it easily.

Neutron Stars

If a supernova is the visible flash of an exploding star, if it exhibits a power far greater than that of an ordinary nova, it would be a logical conclusion, on the basis of the beliefs of the 1920s, that the portion of the star that was not driven out into space as a cloud of dust and gas would collapse into a white dwarf.

The central star of the Crab nebula was hot and bluish, and such a star also existed at the center of the Gum nebula. Perhaps all the other supernova remnants had such white dwarfs at the center that were often too dim to be perceived. It then seemed quite clear that the small, hot stars at the center of the Crab nebula and of the Gum nebula were visible only because those two remnants happened to be relatively close to us.

The first doubt that white dwarfs could be the sole and universal product of stellar collapse arose with the work of the Indian-American astronomer Subrahmanyan Chandrasekhar (1910-   ).

He reasoned that when a star collapsed, the white dwarf that formed no longer had the capacity to undergo fusion reactions so that one could not count on fusion energy to keep it from shrinking.

Yet a white dwarf did not shrink as tightly as it might. If atoms broke up and if matter then shrank until the atomic nuclei were in contact, an object like our Sun would contract to a sphere with a diameter of only about fourteen kilometers (nine miles). White dwarfs were, instead, up to twelve thousand kilometers (7,400 miles) in diameter, and the tiny nuclei were still far enough apart to be able to move about quite freely. And, to be sure, dense as a white dwarf might be, it still behaved, in some ways, like a gas.

Chandrasekhar was able to show that what kept a white dwarf distended was its content of electrons. The electrons no longer existed as parts of atoms but moved about randomly as a kind of electron gas. These electrons repelled one another, and even the intense gravitational field of a white dwarf could not compress the electron gas beyond a certain point.

The more massive the white dwarf, the more intense the gravitational field; and the more intense the gravitational field, the more tightly the electron gas was compressed. It followed that the more massive the white dwarf, the smaller its diameter.

At some point, the ability of the electron gas to resist compression would break down, and the white dwarf would collapse. In 1931, Chandrasekhar calculated that the breakdown would take place at a mass equal to 1.44 times that of the Sun. This is known as "Chandrasekhar's limit."

As it happens, all the white dwarfs whose masses have been determined contain, without exception, masses that are less than 1.44 times that of the Sun.

This did not at first strike astronomers as a problem. Over 95 percent of the stars that exist have masses below Chandrasekhar's limit to begin with, and have no choice, so to speak, but to collapse into white dwarfs.

Then, even for the small minority of stars that have masses above the limit, there seems to be no problem. Before collapsing, stars tend to explode and drive off their outer layers and, hence, lose mass. The more massive the star, the more forceful the explosion and the greater the lost mass. The Crab nebula, comprised of mass lost by the exploding supernova of 1054, has a mass three times that of the Sun.

It was possible to argue that every massive star, before collapsing, would explode and blow off so much of its own mass that what was left of the intact core would always be less than 1.44 times the mass of the Sun and would therefore collapse into a white dwarf.

Yet Chandrasekhar had set up a doubt. What if a star was so massive to begin with that even after it blew away all the mass it could, what was left was still more than 1.4 times the mass of the Sun? In that case, when it collapsed, it would not form a white dwarf. What would happen?

Suppose we reason it out. A white dwarf consists of atomic nuclei and electrons. The atomic nuclei are made up of protons and neutrons. The neutrons have no electric charge, while the protons have a positive electric charge, one that is precisely equal on all protons and is arbitrarily set at unity. Each proton has a charge of +1, in other words.

All electrons also have identical electric charge, but theirs are negative charges. Each electron has a charge precisely counter to that of a proton, so that its charge is −1.

Protons and electrons, having opposite charges, attract each other, but only within limits. When they approach each other too closely, other considerations take over, and a repulsion exists that is far stronger than the attraction of opposite charges. This is another reason, and an even stronger one than the mutual repulsion of the electrons, that keeps a white dwarf from contracting beyond a certain point.

As the gravitational field becomes more intense, however, electrons are pushed closer and closer to one another and to the protons until, at a certain point, the electrons are forced to combine with the protons. When that happens, the opposite electric charges cancel each other. Instead of a negative electron and a positive proton, you get an electrically uncharged combination of the two. In short, you get a neutron.

If a collapsing star has a mass greater than Chandrasekhar's limit, then, as it collapses, the electrons and protons combine to form neutrons, which add to the neutrons already in existence. The collapsing star consists of nothing but neutrons, which, being uncharged, do not repel each other in any way. The star then shrinks until the neutrons are in contact and we have a *neutron star*.

A neutron star can squeeze all the mass of the Sun, as I said earlier, into a ball no more than fourteen kilometers

The tininess of a neutron star compared to the Moon.

(nine miles) in diameter. It is a far smaller star than a white dwarf, far denser, and has a far more intense gravitational field.

In 1934, Zwicky, who was beginning his study of supernovas in other galaxies, speculated about the possible existence of neutron stars as the end-product of the gigantic explosions.

He felt that a supernova, with a million times the energy production of an ordinary nova, was clearly undergoing a far more enormous explosion. The greater explosion should lead to a more catastrophic collapse. Even if the contracting remnant were insufficiently massive to make a white dwarf impossible, it might contract with sufficient rapidity for inertia to carry it through and past the white dwarf stage. For this reason, a neutron star might end up with a mass less than 1.44 times that of the Sun.

It was not long afterward that the American physicist J. Robert Oppenheimer (1904-1967) and a student of his, George Michael Volkoff, worked out the mathematical details of neutron stars and their formation. A Soviet physicist, Lev Davidovich Landau (1908-1968), did the same thing independently.

In the 1930s, then, it seemed quite logical to think that supernovas resulted in the formation of neutron stars, but there seemed no particular way to test the matter by actual observation. Even if neutron stars actually existed, their tiny size would make it certain, it would seem, that even a relatively nearby one, seen in a large telescope, would still be excessively faint. If it could be seen at all, there would be no way of determining anything about it except that it was excessively faint. Thus, the star at the center of the Crab nebula was faint, but in what way could one decide that it was a neutron star, rather than a white dwarf? If anything, just the fact that it could be seen at all seemed to weigh in favor of the white dwarf.

Yet there was one wild hope. The act of catastrophic compression would be accompanied, inevitably, by an enormous rise in temperature so that the surface of a neutron star would, at the time of formation, have a temperature of as much as 10,000,000° C. At such a temperature, even allowing for some thousands of years of cooling, its radiation would include copious quantities of x-rays.

It follows that if a star is small and dim, but x-rays seem to be coming from its position in the sky, there would be a strong suspicion that it was a neutron star.

That one wild hope, however, is accompanied by one sad fact. X-rays do not penetrate the atmosphere—they interact with the atoms and molecules in air and do not survive to reach the Earth's surface as x-rays. Neutron stars might be sending out energetic signals, therefore, and it would not help at all—or, at least, so it seemed in the 1930s.

X-rays and Radio Waves

Of course, if scientists were able to make observations from outside Earth's atmosphere, everything would change.

The one apparent way of reaching beyond the atmosphere was by use of rockets. This had been pointed out in 1687 by Newton. Between knowing this and actually being able to put rockets to practical use, there was a huge gap.

Yet the time did come. During World War II, the Germans made rapid advances in the use of rocket-driven vehicles, owing to the work of Wernher von Braun (1912-1977). Their intention was to make use of them as war weapons, and they succeeded; but, fortunately for the Allies, it was too late in the war. The Germans lacked the time needed to deploy them in sufficient quantities to stave off defeat.

After the war, however, both the United States and the Soviet Union took up rocket research where the Germans had left off. In 1949, the United States succeeded in sending rockets high enough to have them reach effectively beyond the atmosphere, and, in 1957, the Soviet Union actually put a rocket-propelled object into orbit about the Earth.

Now it became possible to detect x-rays from space, and at once certain problems could be solved.

Thus, the spectrum of the Sun's corona (its outer atmosphere) had spectral lines that could not be identified with those produced by known elements. Some, therefore, speculated that a hitherto unknown element, "coronium," existed in the corona.

In 1940, on the other hand, the Swedish physicist Bengt Edlen (1906-   ) maintained that these lines represented the atoms of known elements that existed in very unusual states because the corona was at a high temperature of 1,000,000° C. or more.

How could one check, then, on whether coronium existed or not? If Edlen were right, then there should be x-rays emitted in quantity from the super-hot corona, but in 1940 there was no way of detecting those x-rays, even if they existed.

Once rockets were available, things changed. In 1958, the American astronomer Herbert Friedman (1916-   ) supervised the firing of six rockets that would rise above the atmosphere and be capable of detecting x-rays from the Sun, if such existed. The x-rays *were* detected; the corona was as hot as Edlen had suggested; the spectral lines were indeed those of ordinary elements under very unusual conditions—and coronium did *not* exist.

The Sun's emission of x-rays is but mild, however. Such x-rays are easy to pick up only because the Sun is so close to us. Even the nearest stars, those of the Alpha Centauri system, are 270,000 times as far from us as is the Sun. If one of the stars of the Alpha Centauri system produced x-rays as intensely as the Sun did, we would receive only 1/70,000,000,000 as intense a beam of x-rays from it as from the Sun, and we would not detect that beam. X-rays from stars that were still farther away would be even less likely to be detectable.

It follows that if the universe consists only of Sun-like stars, it would be very unlikely, with detection systems of the type available to us now, that we would detect any x-ray source in the sky, other than the Sun itself. If, on the other hand, there were unusual stars that gave off enormous intensities of x-rays—as neutron stars might—they *would* be detected.

It became extremely important, then, to try to determine what x-ray sources, if any, might be in the sky, for every x-ray source bore the promise of signifying something unusual.

In 1963, Friedman detected nonsolar x-ray sources in the sky, and, in the years since, numerous such sources have been detected. In 1969, for instance, a satellite was sent up that was expressly designed for the detection of x-ray sources. It was launched from the coast of Kenya on the fifth anniversary of Kenyan independence and was named "Uhuru," from the Swahili word for "freedom." It detected no fewer than 161 x-ray sources, half of them from outside our galaxy.

This was one of the ways in which, in the 1960s, astronomers began to realize that the universe was a much more violent place than had earlier been suspected. The apparent calm and serenity of the night sky was misleading.

One of the x-ray sources in the sky was the Crab nebula.

This came as no surprise to astronomers. If they had to pick one spot in the sky from which there would be detectable x-rays, every last one of them would undoubtedly have picked the Crab nebula. For one thing, it was the certain remnant of a supernova explosion, the most catastrophic event that could possibly involve a star. Also, its explosion was reasonably close and reasonably recent. What's more, the enormous turbulence and rapid expansion of the nebula gave every promise of the kind of high temperature that would produce x-rays.

There were indeed two possible sources of x-ray emission. One was the rapidly expanding volume of gas and dust that made up the nebula proper. The other was the small, hot star at the center, the remnant that *might* be a neutron star.

As it happens, the Moon, in its movement across the sky, was going to cut across the Crab nebula in 1964. Little by little, it would encroach on the nebulosity.

If the x-rays were being produced by the hot turbulent gases of the nebula itself, the x-ray intensity would be cut gradually as the Moon eclipsed it. If the x-rays were being produced chiefly by the supposed neutron star in the center, then the intensity would decline as the Moon moved in front of the nebula, then drop sharply as it passed the tiny star, then resume the slow decline as the rest of the nebula was eclipsed.

When the time for the eclipse came, an x-ray detecting rocket was sent up, and, from the observed results, it appeared that the intensity declined regularly. There was no clear indication of a sudden drop. The hopes for the detection of a neutron star withered.

Yet they did not die out altogether. The mere fact that both the central star and the surrounding gases could each serve as an x-ray source introduced the possibility of confusion. If it were only possible to find something that would characterize the star itself and not the surrounding gases, the riddle might be read.

But what might that something be? When the answer came, it came completely unexpectedly.

X-rays and gamma rays are at the energetic end of the electromagnetic spectrum. At the other, low-energy end are the radio waves.

Radio waves do not, in general, penetrate the atmosphere any more than x-rays do. In the case of radio waves, the problem is a layer of the upper atmosphere that is rich in electrically charged particles, the ionosphere. The ionosphere tends to reflect radio waves so that those that originate on Earth and radiate upward are reflected back to Earth. In the same way, radio waves that originate from astronomical objects would be reflected back into space by the ionosphere and would never reach the Earth's surface.

This is not true, however, for a stretch of the shortest radio waves, the microwaves. The wavelength of the microwaves is very short for radio waves ("micro" is from a Greek word for "short"), but it is much longer than that of ordinary light waves or even than the radiation of the infrared region.

What it amounts to, then, is that in the electromagnetic spectrum there are two regions where radiation can pass through Earth's atmosphere with little loss. One is the visible light region and the other is the microwave region, the latter being the broader.

We have known of the "light window" as long as we have existed, for we have eyes that can sense light, and we can see the Sun, Moon, planets, and stars. We do not detect the "microwave window" through any natural sense organ, however, and it is only in the last half-century that we have become aware of it.

The microwave window was discovered accidentally by the American radio engineer Karl Guthe Jansky (1905-1950) in 1931. Working for Bell Telephone, he was trying to pinpoint the source of static that interfered with radio reception. In the process, Jansky's receiving device recorded a hiss that came from the sky. It seemed at first to be caused by microwaves coming from the Sun, but, with time, the source moved farther and farther from the Sun and, by 1932, Jansky found the source to be located in the constellation Sagittarius. We know now that it was coming from the center of the galaxy.

Jansky's discovery was not followed up at once by professional astronomers, since the techniques for detecting microwaves were not yet well developed. However, an amateur radio enthusiast, Grote Reber (1911–   ), who heard of Jansky's report, built an elaborate paraboloid detector in his backyard in 1937. (He was only sixteen at the time.) This was the first "radio telescope," and with it Reber scanned the sky to find particular radio sources. In this way he made the first radio map of the sky.

At about the same time, the Scottish physicist Robert Watson-Watt (1892–1973), among others, was helping to perfect a method for detecting the direction and distance of otherwise unseen objects by using a beam of microwaves. The microwaves would be reflected from the object, and the reflection could be detected. The direction from which the reflection came gave the direction of the object, and the time interval between emission of the beam and detection of the reflection gave its distance. The technique was called *radar*.

Radar turned out to be of crucial importance during World War II, and, by the end of the war, satisfactory techniques for emitting and receiving microwaves had been worked out. It meant that after the war, astronomers could study and analyze the microwave emissions from distant star clusters in great detail. Better and better radio telescopes were built, and large numbers of crucial and, for the most part, unexpected discoveries were made as a result. An astronomical revolution took place that matched, in importance, the one produced by the invention of the telescope three and a half centuries before.

Pulsars

In 1964, radio astronomers became aware that radio sources were not necessarily steady, any more than light sources were.

Light waves are refracted by the atmosphere to different extents, according to temperature. Because the atmosphere contains regions of different temperature, and because temperature changes with time, the feeble light issuing from stars is bent this way and that, the direction changing slightly with time, so that the star seems to "twinkle." Radio waves are deflected by the charged particles of the atmosphere this way and that, randomly, so that radio sources also seem to twinkle.

In order to study this rapid twinkling, or *scintillation*, specially designed radio telescopes had to be constructed, and one such was devised by an English astronomer, Antony Hewish (1924-   ). It consisted of 2,048 separate receiving devices spread out over an area of 18,000 square meters.

In July 1967, Hewish's radio telescope began scanning the sky in order to detect and study the twinkling of radio sources. At the controls was a student of his, the English radio astronomer Susan Jocelyn Bell (1943-   ).

In August, Bell noticed something peculiar. There was marked scintillation from a particular source between the stars Vega and Altair that was observed at midnight, a time when scintillation was usually low. What's more, the scintillation seemed to come and go. She brought this to Hewish's attention, and, by November, it seemed worth concentrating on.

The radio telescope was adjusted to make a high-speed recording, and it turned out that superimposed on the scintillation was an occasional burst of radiation that was very brief, lasting only one-twentieth of a second. That was why the scintillation seemed to come and go. When the source was not being observed very closely, the instrument scanning it occasionally passed over it just as the burst of radiation came, but usually it did so between bursts.

117

As the bursts continued to be studied, it was discovered that they came at brief and regular intervals, *very* regular. The interval between bursts was about 1⅓ seconds long, or, to take it to eight decimal places, the bursts came at intervals of 1.33730109 seconds.

Nothing in the sky had ever been observed to take place so regularly and at such brief intervals. Whatever it was that caused it, it was unprecedented. It had to be something cyclic; it had to be an astronomical object that was revolving about another object, or rotating about its axis, or pulsating, and, for some reason, giving off a burst of microwaves at each revolution, or rotation, or pulsation.

Pulsation seemed the best bet at first, and Hewish called it "a pulsating star," a phrase that was quickly abbreviated to *pulsar*.

Once Hewish knew the manner in which a pulsar emitted microwaves, such objects became easy to detect. Each pulse produced an intense enough burst of microwaves. The trouble was, though, that ordinary radio telescopes didn't pick up the individual burst but only the average emission spread out over a period of time. If the bursts were averaged together with the quietness of the inter-burst periods, the level of microwave intensity was only about one-twenty-seventh that of the burst peak, and this average is not high enough to be particularly noticeable.

Hewish's radio telescope could detect the bursts, and he began to comb the sky, looking for more of the same. By February 1968, three more pulsars had been discovered, and Hewish then felt safe in announcing the discovery.

At once, others took up the search, and, quickly, five more pulsars were found. By the early 1980s, nearly 400 pulsars had been identified.

One pulsar was discovered in October 1968, in a place where anything strange might be expected to exist—the Crab nebula. It proved to have a much more rapid pulsation than the first one. Its period is only 0.033099 seconds, meaning that the microwave bursts emerge about thirty times a second. Another pulsar was later discovered at the center of the Gum nebula.

Here there was no chance of confusion. If it were simply a matter of the steady emission of radiation, whether x-rays or radio waves, it might be hard to disentangle the portion coming from the central star from the portion coming from the nebulosity. The very rapid and regular pulsation, however, could be located precisely, for it came from one spot only and was not emanating from an area. And that one spot coincided with the central star in the case of the Crab nebula as well as the Gum nebula.

The understanding arose that just as the central star of a planetary nebula is a white dwarf, the central star of a supernova remnant is a pulsar. To put it another way, a star that explodes into a supernova collapses into a pulsar.

But what is a pulsar?

The short period of the microwave pulses show that a pulsar must be pulsating, rotating, or revolving in no more than a few seconds, and sometimes in but a small fraction of a second. No object can undergo any sort of cyclical change so rapidly unless it is very small and has a very intense gravitational field to keep it from breaking up under the inertial stresses of such rapid cycling.

One known object that is both small in size and intense in gravitational field is a white dwarf—but even a white dwarf is not small enough nor does it have an intense enough gravitational field. There seemed nothing to do but to suppose that a pulsar was a neutron star. That, at least, would be small enough and would have an intense enough gravitational field.

It didn't seem likely that a neutron star, with its unimaginably intense gravitational field, could pulsate. Nor could a neutron star revolve about any object (even another neutron star) in a fraction of a second. By elimination, only one thing was left, and that was a rotating neutron star. A neutron star could, in theory, rotate not only thirty times a second as the Crab nebula pulsar would require but up to a thousand times a second and more. In November 1982, a pulsar was discovered that emitted bursts of microwaves 640 times a second, so that it seemed to be a neutron star rotating about its axis in only a little over $\frac{1}{1,000}$ of a second. It is called the "millisecond pulsar."

But why should a rotating neutron star send out bursts of microwaves?

A number of astronomers, including the Austrian-born Thomas Gold (1920-   ), studied the problem. They argued that such an extremely condensed star would have an enormously intense magnetic field, and the magnetic lines would spiral around and around the rapidly spinning neutron star.

Consider the extraordinarily high temperature of a neutron star. It would be expected to give off speeding electrons, the only objects that would move fast enough to escape from its surface against the intense gravitational pull. Since the electrons are electrically charged, they would be trapped by the magnetic lines of force and be able to escape only at the magnetic poles of the neutron star. These magnetic poles would be at opposite sides of the star, but not necessarily at the rotational poles. (Earth's magnetic poles are quite far from the rotational poles, for instance.)

As the electrons move away from the neutron star, following the sharply curved path enforced upon them

Rotating neutron stars send out a double stream of microwaves, which we can sometimes detect.

by the magnetic lines of force, they lose energy in the form of a spray of radiation—microwaves, among other things. As the neutron star rotates, one, or sometimes both, of the magnetic poles may move across the line of sight to Earth, and we would receive a spray of microwaves each time that happened. Thus, the rotating neutron star pulses. The faster the rotation, the faster the pulse.

Since the radiation, which is given off as the escaping electrons lose energy, should be across the whole range of the electromagnetic spectrum, we ought to be receiving pulses of light from rotating neutron stars as well as of microwaves.

We can see the pulsar at the center of the Crab nebula, however, and its light seems to be steady. But since it ought to be flickering thirty times a second, we should expect to see it steady, just as we see the motion picture screen in continuous motion even though we are actually seeing a succession of movie "stills" projected at a rate of sixteen per second.

In January 1969, three months after the Crab nebula pulsar was first detected, its light was studied stroboscopically; that is, its light was permitted to pass through a slot that was open only for about a thirtieth of a second. Once that was done and the star was photographed in very short intervals of time, it was found that there were short intervals when it was on and short intervals when it was off. It flicked on and off, thirty times a second. It was an "optical pulsar."

Gold went on to point out that, if the identification of pulsars with rotating neutron stars were correct, then these neutron stars were losing energy steadily and the rate of rotation should slowly decrease. The pulse of radiation should come at gradually increasing intervals. The change would be exceedingly tiny, but the pulses were so regular that even exceedingly tiny changes should be measurable.

Thus, when the Crab nebula pulsar was first formed 900 years ago at the time of the supernova explosion, it may have been rotating on its axis 1,000 times a second. It would have lost energy quickly, and, in the first 900 years of its existence, over 97 percent of its energy has likely bled away until it is now rotating only thirty times a second. The period of rotation should still be slowing— though, of course, more and more gradually.

To check Gold's suggestion, the period of the Crab

nebula pulsar was studied carefully and its rotation was indeed found to be slowing. The interval between pulses is increasing by 36.48 billionths of a second each day, and, at that rate, the interval will have doubled in 1,200 years.

The same phenomenon has been discovered in other pulsars whose periods are slower than that of the Crab nebula pulsar and whose rate of slowing is therefore also slower. The first pulsar discovered, which has a period that is forty times as long as that of the Crab nebula pulsar, is slowing at a rate that will double its period after sixteen million years.

As a pulsar slows its rotation and lengthens its pulsation period, the pulses become less energetic. By the time the period has passed four seconds in length, the individual pulses are not sufficiently more intense than the general background of the universe to be easily detectable. Pulsars probably endure as detectable objects for three or four million years, however.

But there is one finding that doesn't fit in with the neat pattern just described. The millisecond pulsar recently discovered, which I mentioned earlier, rotates in a little over $\frac{1}{1,000}$ of a second and therefore ought to be very young. Yet its other properties identify it as actually a very old pulsar. What is more, its period does not seem to be lengthening noticeably.

Why should that be? What keeps it spinning so rapidly? The most reasonable suggestion at this time is that such a pulsar gains mass from a nearby companion star in such a way that its spin is accelerated.

6

KINDS OF EXPLOSIONS

Types I and II

It might seem astonishing, and even gratifying, that in the space of fifteen years, astronomers should have discovered nearly 400 stars of a type whose existence had been unknown prior to an accidental discovery in 1969. And yet, viewed in another fashion, the question we might ask is: Why so few?

Suppose that neutron stars are the inevitable remnants of supernovas and that supernovas explode in our Milky Way galaxy at the rate of one every fifty years. In that case, if we suppose our galaxy to have been in existence for fourteen billion years, and the rate of supernova explosion to have been constant throughout that time, then the total number of supernova explosions would have been 280 million. Would that not mean we ought to expect to find that many neutron stars, or about one for every 900 stars in the galaxy? Why then only 400 altogether?

Let's think about it, though. It doesn't matter how many billions of years the Milky Way galaxy has existed, if neutron stars remain detectable for only four million years or so. In that case, the vast majority of neutron stars that may exist would be too old to detect, and only those that formed in the last four million years could possibly be sending out pulses of radiation strong enough to register on our instruments.

If we confined ourselves to the last four million years, then, we would be dealing with 80,000 supernovas and therefore, at most, 80,000 potentially detectable neutron stars in the Milky Way galaxy today.

To be sure, only a minority of those 80,000 supernovas would have been visible from Earth, a majority being hidden by interstellar dust clouds. It is, however, *light* that is hidden. Radio waves penetrate the dust clouds with ease, and that means that the spray of microwaves sent out by pulsars can be detected by our radio telescopes even in cases where the original supernova would have been hidden from our optical telescopes.

But who is to say that the spray of microwaves will be in our direction? It is quite possible that a neutron star, in its rotation, sprays microwaves and other radiation in a circle that at no point touches Earth. We could not possibly detect such a neutron star, however energetic it might be, by any present-day technique.

If we consider the number of neutron stars that might exist that are less than four million years old and that happen to be spraying in our direction, the total might sink to 1,000 or so (although some of the more optimistic astronomers make the figure much higher).

We must also take into account the fact that not

every supernova necessarily produces a neutron star, and this would lower the figure of detectable neutron stars even further. It might even seem (though this may be unnecessarily pessimistic) that we are now approaching the limit of the number of neutron stars we might find.

In the survey of supernovas in our galaxies that began with Zwicky's work in the 1930s, astronomers have learned to distinguish among them through differences in light curves and other properties. It is generally accepted, now, that there are two types of supernovas, which are usually labeled Type I and Type II.

Type I supernovas tend to be the more luminous of the two, reaching an absolute magnitude of as much as −18.6, or 2.5 billion times the luminosity of our Sun. If such a supernova were at the distance of Alpha Centauri, it would appear, at peak brilliance, about one-seventh as bright as the Sun. Type II supernovas are a bit dimmer, brightening to only about one billion times the luminosity of the Sun.

A second difference is that Type I supernovas, having reached and passed their peak brilliance, decline in brightness in very regular fashion, while Type II supernovas do so much more irregularly.

A third difference is obtained from a study of the light spectra. Type I supernovas seem to show an almost total lack of hydrogen, while Type II supernovas are rich in hydrogen.

A fourth difference rests with the location. Type II supernovas are found almost always in spiral galaxies and, what's more, in the arms of those galaxies. Type I supernovas are much more general in their location preferences, exploding not only in spiral arms but in the central sections of spiral galaxies and in elliptical galaxies as well.

The difference in location tells us something important at once. Elliptical galaxies are largely dust-free. Their stars are, on the whole, relatively small stars, just a little larger than our Sun at most, and have existed for most or all of the life of the galaxy. The same is true of the central regions of spiral galaxies.

The arms of the spiral galaxies are, however, dust-laden, and as we shall see later on, they are the site of many young and massive stars.

Type I supernovas, then, must involve stars that possess about the mass of the Sun or a little more. Type II supernovas must involve stars that are considerably more massive than the Sun, at least three times as massive and perhaps, in some cases, much more than that.

The more massive a star, the less common it is. The relatively small stars involved in Type I supernovas are at least ten times as common as the massive ones involved in Type II supernovas, and one might expect, therefore, that Type I supernovas are at least ten times as common as Type II supernovas.

Not so! The two are equally common. From this we can deduce that not every small star will end up as a Type I supernova; that, in fact, only a small minority will. The requirements for Type I supernovas are more stringent, then, than we might expect. Not only is a roughly Sun-size star required, but a special type of star of this size.

Here we can turn to the chemical differences between the two types of supernova. The Type I supernovas have virtually no hydrogen, which means they are at the latter end of their evolutionary development. In fact, if a star has no hydrogen and is rich, instead, in carbon, oxygen, and neon, we would feel safe in saying that it is a white dwarf. We conclude, then, that the Type I supernovas must represent exploding white dwarfs.

Left to themselves, white dwarfs do not explode and are quite stable. As we already know, however, white dwarfs are not always left to themselves. They are sometimes part of a close binary star system. In that case, when the companion star of the white dwarf, in the course of *its* evolution, swells to a red giant, matter will spill over into an accretion disk that periodically adds mass to the white dwarf.

We have already seen that, periodically, the matter added to the white dwarf will be heated and compressed to the point of undergoing fusion. There is a vast explosion, what is left of the accretion disk is driven away, and the white dwarf greatly multiplies its luminosity (temporarily) and is seen from Earth as a nova. This will repeat itself at longer or shorter intervals.

With each episode of nova formation, some of the mass of the accretion disk will be held on to by the white dwarf so that its overall mass will gradually increase.

But what if the white dwarf is particularly massive for such an object and is, say, 1.3 times the mass of our Sun? Or what if the companion star is unusually massive and expands into an unusually large red giant so that it will spill mass over into the white dwarf's gravitational influence at a much more rapid than average rate? Or suppose both things are true.

In such cases, the white dwarf may, fairly rapidly, gain enough mass to push it over Chandrasekhar's limit, which is 1.44 times the mass of our Sun. Once that happens, the white dwarf cannot maintain itself as such.

The white dwarf, instead, collapses and caves in. It compresses very rapidly and slams the nuclei of carbon and oxygen together with great force. All of it undergoes fusion at once, producing so much energy so rapidly that the result is a vast explosion that radiates as much energy in a few weeks as our Sun will produce in all its multi-billion-year lifetime. In short, the collapse of the white dwarf and the fusion of its substance produces not just a nova but a Type I supernova.

Such a Type I explosion tears the star apart and may leave behind no collapsed star at all of any kind—no white dwarf, no neutron star, but only a turbulent and expanding cloud of dust and gas. Tycho's nova of 1572 and Kepler's nova of 1604 were both, in all likelihood, Type I supernovas, and in neither case has a neutron star been detected at their sites—only nebulosities.

Type II supernovas also take place at the latter end of a star's evolution, but at a stage not quite as far along as in the case of a Type I supernova. The Type II supernova occurs in a star that has reached the red-giant stage. However, it occurs in a massive star, one that is at least three to four times as massive as our Sun, and the more massive a star the larger the red giant.

A really large red giant consists of various layers, like an onion. The outermost layer is still hydrogen and helium, the mixture that makes up most of a normal star on the main sequence. Under that is a shell containing the nuclei of more massive atoms, such as those of carbon, nitrogen, oxygen, and neon. Under that is a third shell, rich in the nuclei of sodium, aluminum, and magnesium. Under that is a fourth shell, rich in the nuclei of sulfur, chlorine, argon, and potassium. And at the core is a fifth shell, rich in the nuclei of iron, cobalt, and nickel.

Each shell below the outermost is composed of the product of the fusion of the smaller nuclei that still exist in the shell outside. Once a star develops a core of iron, cobalt, and nickel, it can go no further. Any additional nuclear change involving these nuclei, whether fusion into more complicated nuclei or fission into less complicated nuclei, will not *release* energy but will *absorb* it instead.

As the iron core grows larger, the star reaches a stage where, as a whole, it cannot release enough energy to keep itself extended. The inner layers contract catastrophically, and the gravitational energy so released explodes the outer layers outward and, in addition, induces fusion within them, releasing still more energy. This energy is what makes itself evident as a Type II supernova, and it brings about even those nuclear reactions that absorb energy.

The collapsed core of such a supernova is quite likely to form a neutron star even when the mass (after the exploded outer layers of the star are subtracted) is small enough to allow a white dwarf to exist. The collapse is so catastrophic that the core plunges through the white-dwarf level, so to speak, without stopping.

Black Holes

Even in the case of a Type II supernova, it is not inevitable that a neutron star be formed.

In 1939, when Oppenheimer was working out the theoretical implications of the neutron star, he studied the possible consequences of increasing the mass of the star. Naturally, as the mass increases, the intensity of the star's gravitational field also increases. When the mass becomes greater than 3.2 times that of our Sun, the gravitational field becomes so intense that even neutrons in contact cannot withstand the compression induced by the field. The neutrons collapse and the neutron star contracts and grows steadily more dense—which means that, in consequence, the gravitational field in the neighborhood of the tiny star becomes still more intense, and the contraction continues still more rapidly.

Once the neutrons collapse, there exists no known way in which the contraction can be stopped. So it seemed to Oppenheimer at the time, and so it still seems to scientists today. The only conclusion is that the compression continues indefinitely so that the star would approach zero volume and infinite density.

This does not mean that we are merely dealing with smaller and smaller, and denser and denser neutron stars. As the contraction continues, an important change takes place:

In order to see the nature of the change, let's imagine an object being thrown upward from the surface of the Earth. As it moves upward, Earth's gravitational field pulls downward upon it steadily. Its upward speed decreases steadily as a result. The object is finally brought to a standstill, and, in the next instant, it begins to fall.

If Earth's gravitational field were equally intense all the way upward, this would happen no matter how rapid the object's initial upward speed would be. Eventually, after 100 meters, or 100 kilometers, or 100,000 kilometers, that speed would be slowed to zero, and the object would begin to fall and would return to Earth.

Earth's gravitational field is, however, *not* equally intense all the way up but diminishes as the square of the distance from Earth's center.

At Earth's surface, an object is 6,370 kilometers (3,950 miles) from the center. At a height of 6,370 kilometers above the surface, the distance from the center has doubled and the intensity of Earth's gravitational field has been reduced to one-quarter of what it was upon its surface. It continues to decrease in this way with increasing height. At the distance of the Moon, the intensity of Earth's gravitational field is only 1/3,500 that at its surface.

If an object is thrown upward at sufficient speed, it can, so to speak, outpace the gravitational field. The field will act to slow it down, but the field will weaken so quickly, as the object moves swiftly upward, that the steadily diminishing gravitational pull can never manage to reduce the upward motion to zero. The object can, in this way, escape the Earth's gravitational field and wander through the universe indefinitely. Of course, it may still be in the grip of other, more massive objects than the Earth—such as the Sun—or it may encounter another body in its wanderings and collide with it or go into orbit about it.

The minimum speed at which a moving object at Earth's surface can just barely escape Earth's gravitational field is the *escape velocity*. For Earth, the escape velocity is 11.2 kilometers (6.9 miles) per second.

A more massive object, with a consequently more intense gravitational field, will naturally require a higher escape velocity at its surface. For Jupiter, the value is 60.5 kilometers (37.5 miles) per second, and for the Sun it is 617 kilometers (383 miles) per second.

If a star contracts, the gravitational field at its surface grows more intense as that surface approaches the center, even though the total mass of the star may not change. Thus, Sirius B, the first white dwarf to be investigated by astronomers, has a mass roughly equal to that of the Sun, but its surface is much closer to its center than is true for the Sun. The surface gravity of Sirius B is therefore much more intense than that of the Sun, and the escape velocity from the surface of Sirius B is about 4,900 kilometers (3,038 miles) per second.

The higher the escape velocity from an astronomical body, the more difficult it is for anything to escape from that body, and the less likely it is that anything will, in actual fact, do so.

In the last quarter-century, our rockets have attained speeds great enough to make escape from Earth's gravitational field possible, but if our planet's surface gravity were somehow to be increased to that of Jupiter (without inconveniencing us personally), our technological expertise would no longer suffice to send rockets into outer space.

A neutron star with the mass of the Sun would have

an escape velocity of about 200,000 kilometers (124,000 miles) per second. At that point, not only would our present-day technology fall short of getting anything off such an object but almost anything would. The only objects that would normally move rapidly enough to escape from the surface of a neutron star would be very energetic particles of low mass or particles of no mass at all. Energetic electrons could escape, and so could neutrinos or the photons that make up light and similar radiation.

If a neutron star collapses, the gravitational intensity continues to increase without limit, and the escape velocity continues to rise. At a certain point, the escape velocity reaches the mark of 300,000 kilometers (186,000 miles) per second. That happens to be the speed of light in a vacuum, and, as the German-born scientist Albert Einstein (1879-1955) argued in 1905, that is the fastest conceivable speed. Nothing with mass can reach that speed, and even massless particles, while traveling at that speed, cannot exceed it.

This means that when the collapsing neutron star reaches that stage, nothing can leave it (except under very special circumstances that need not concern us here). Anything that collides with it behaves as if it had fallen into an infinitely deep hole from which it can never again emerge. Even light cannot escape it. The American physicist John Archibald Wheeler (1911-   ) used the term *black hole* to describe it. The name caught on at once.

It follows, then, that if the contracting core of a supernova has a mass of more than 3.2 times that of our Sun, it smashes right through the white dwarf and neutron star stages and ends up as a black hole.

Thus, a Type II supernova, though it often produces a neutron star, will also often produce a black hole. Consequently, since neutron stars are produced by only one type of supernova and, even then, not always, we need not be surprised that there are fewer pulsars than the number of supernovas might lead us to expect.

There is an important practical difference between black holes and neutron stars—black holes are almost impossible to detect.

We can detect a neutron star, easily enough, by the sprays of radiation it emits. But a black hole emits nothing to speak of, not even radiation. The ordinary techniques by which we detect other astronomical objects simply won't work for isolated black holes.

An isolated black hole could only be detected by us if it were massive enough, or close enough, or both, to affect us gravitationally. There could, in theory, be millions of black holes scattered throughout the galaxy, each with the mass of an ordinary star, and we could well remain unaware of the fact.

Yet radiation could originate in the neighborhood of a black hole if not from the object itself. A black hole is never truly isolated. There is always matter in its vicinity, even if only the thin wisps of atoms and dust that exist in interstellar space. Matter that approaches a black hole, even in the form of the occasional bit, can move into an accretion disk about it. Little by little, such matter will spiral into the black hole and emit synchrotron radiation in the form of x-rays.

The x-rays emitted by a black hole surrounded only by interstellar matter are, however, so low in intensity that they could barely be detected, if at all, and would give us no useful information.

Suppose, though, that a black hole is near a large source of matter, so that large masses are constantly spiraling into it and, in the process, giving off intense x-ray emissions. This would take place if we were dealing with a close binary system, the sort of thing that would produce novas, or even Type I supernovas, if one partner were a white dwarf.

If one partner were a black hole, there would be no question of an explosion, of course. The black hole would merely grow more massive as it absorbed matter, for there are no upward limits to the mass of a black hole. However, as the black hole grows, x-rays from the infalling matter would be continually emitted from a point where, otherwise, nothing could be seen.

For that reason, astronomers grew interested in x-ray sources.

In 1971, the x-ray detecting satellite Uhuru showed that a strong x-ray source, in the constellation Cygnus, varied irregularly, which seemed to eliminate it as a neutron star and to raise the possibility of a black hole.

Attention was focused on the source, and microwave emission was detected and pinpointed very accurately. The source of emission was very close to a visible star, listed in the catalogs as HD-226868. This is a very large, hot, bluish star about thirty times as massive as our Sun. On close examination, this star proved to be a binary, circling in an orbit with a period of 5.6 days. From the nature of the orbit, the other member of the binary would seem to be five to eight times as massive as the Sun.

The companion star cannot be seen, however, even though it is an intense source of x-rays. If it cannot be seen, it must be very small. Since it is too massive to be either a white dwarf or a neutron star, the inference seems to be that the invisible star is a black hole.

Furthermore, HD-226868 seems to be expanding as though it were entering the red-giant stage. Its matter would therefore very likely be spilling over into the black-hole companion, and it would be the accretion disk about the black hole that would be producing the x-rays.

135

Assuming that the companion of HD-226868 is a black hole (and the evidence is still indirect), then it is undoubtedly the remnant of an ancient supernova.

The Expanding Universe

Although supernovas are magnificent explosions, far beyond anything we can imagine, they are not the greatest explosions that have ever existed. There are some "active galaxies" in which the entire core seems to be exploding, producing far more energy over a far longer period than supernovas can. And we can even go beyond that.

What's more, we *must* do so, for only then can we begin to consider what effect supernovas may have on us.

Do supernovas have any effect on us, we might ask? Can they?

It might at first seem that they needn't really concern us at all, in any practical sense. Only a small fraction of the stars that exist ever explode as novas or supernovas, and, for the foreseeable future, we know of no star near us that is likely to do so.

If our Sun were itself a star that might go nova or supernova someday, then that could well focus our attention on the process with a sort of grisly fascination—but our Sun is safe. It is not massive enough ever to explode as a Type II supernova; and it is not a member of a close binary system, so that it will never be a Type I supernova—or even an ordinary garden-variety nova.

It is, in fact, quite possible to argue that no star that is capable of going nova or supernova could ever be accompanied by a planet upon which intelligent life existed.

If a star were sufficiently massive to form a Type II supernova eventually, then it is easy to argue that it would be too massive to last long enough on the main sequence for life to evolve to the point of producing intelligent beings.

If, on the other hand, it were no more massive than the Sun but were a member of a close binary, so that it might someday explode as a nova or as a Type I supernova, it might not be possible for a planetary orbit to exist about the binary that would provide a stable enough environment for life to develop.

So what, then, have novas and supernovas to do with us? Can't we say that, barring a very occasional side glance at some fairly bright star in the sky, we get nothing out of them either for good or for evil, and should leave them strictly to astronomers and science fiction writers?

We could come to such a conclusion, indeed, but only if we are completely uninterested in how our universe was formed, how the Sun and Earth came to be, how life has evolved, and what dangers may possibly face us in the future—for exploding stars play an intimate role in every one of these things.

To begin with, how was the universe formed?

Until quite recent times, it was taken for granted by most (if not all) cultures, certainly including our own, that the universe was formed over a short period of time, not very long ago, by the magical action of a supernatural being.

In our own culture, the general opinion has supposed the universe to have been formed by God in a period of six days, some six thousand years ago. There is no physical evidence for this, and the belief rests solely on the statements in the first chapter of the Biblical book of Genesis. Still, few people dared to express doubts about the matter even if they had some.

As modern astronomy made it clear that the universe

was enormous, and as every further advance made the universe appear more and more enormous until it seemed incomprehensibly large, it became difficult and, indeed, quite impossible for a rational human being to believe that the Biblical tale of the Creation was literally true.

Yet, on the other hand, there seemed nothing in astronomical observations that could give rise to a purely natural account of the creation.

There was Laplace's nebular hypothesis that gave an interesting and plausible account of the evolution of the solar system from a slowly rotating mass of dust and gas—but where did the dust and gas come from?

Presumably all the stars in the galaxy were thus formed, so there must originally have been a galaxy-sized mass of dust and gas that evolved into many billions of stars and planetary systems. Then, in the 1920s, when it came to be understood that there were innumerable galaxies, that meant there must have been innumerable such masses of dust and gas to begin with. Where did it all come from? How could one possibly work out the origin of huge masses of dust and gas spread over a universe that is billions of parsecs in diameter, without falling back on an omnipotent supernatural being?

However, observations were made in the 1910s that had, apparently, nothing to do with the problem, and that ended by revolutionizing our thinking on the matter.

This began with the American astronomer Vesto Melvin Slipher (1875-1969), who obtained the spectrum of the Andromeda galaxy in 1912 (when it was not yet understood to be a galaxy). From its spectrum, he determined that it was approaching us at the rate of 200 kilometers (124 miles) per second.

He did this by observing that the identifiable dark lines of the spectrum were shifted from their normal position toward the violet end of the spectrum. From the direction of the shift, he could tell that the Andromeda galaxy was approaching us, and from the amount of the shift, he could calculate the speed of approach. This was based on a principle first advanced, in 1842, by the Austrian physicist Johann Christian Doppler (1803–1853).

This "Doppler effect" was applied to sound waves at first, but the French physicist Armand H. L. Fizeau (1819–1896) showed, in 1848, that the principle applied to light waves, too. By the "Doppler-Fizeau effect," it became clear that if the spectral lines of any light-emitting object, whether candle or star, shifted toward the violet, the light source was approaching us. If it shifted toward the red, the light source was receding from us.

The first to apply this principle to a star was William Huggins in 1868. He found that the star Sirius showed a small "red shift" and was therefore receding from us. In the years that followed, other stars were tested in this way. Some were approaching, some receding, with speeds of anywhere up to 100 kilometers (62 miles) per second.

The Doppler-Fizeau effect had one particularly useful aspect. If one attempted to measure the proper motion of a star (motion across the line of sight), success could only be achieved for a star that was quite close. The result is that very few stars have measurable proper motions. The Doppler-Fizeau determination of *radial motion* (toward or away from us) could, on the other hand, work for any star, however distant, provided it was bright enough to yield a spectrum.

Once the Andromeda galaxy was made to yield a spectrum that could be photographed, it didn't matter that it was 700,000 parsecs away (something Slipher had no idea of, to be sure). The Doppler-Fizeau effect worked as well for it as for Sirius—or as for a nearby candle. The

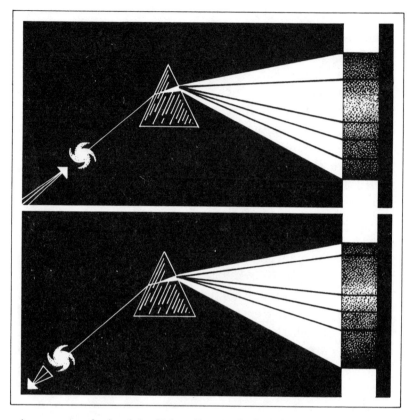

A proper analysis of starlight tells us whether the star is approaching or receding, and how quickly.

"violet shift" in the spectrum of the Andromeda galaxy showed it to be approaching, which was not surprising. The speed of its approach was a little high, since no star had yet been found to be either approaching or receding at such a speed, but, nevertheless, the figure for the Andromeda galaxy wasn't completely out of line.

Slipher went on, thereafter, to study the spectra of fourteen other galaxies (or nebulas, as he thought of them) and found that only one of them was approaching, as the Andromeda was. All the others were receding, and at speeds markedly *greater* than 200 kilometers (124 miles) per second.

This was surprising, indeed, but matters were to grow more surprising still.

In the 1920s, when it came to be realized that the white nebulas were other galaxies, the American astronomer Milton La Salle Humason (1891-1972), working with Hubble, began to photograph the spectra of hundreds of galaxies. What he found was that all, without further exception, showed red shifts: All were receding.

What's more, the dimmer (and therefore, presumably, the more distant) the galaxy, the greater the red shift and the faster the speed of recession. By 1919, Hubble suggested that there was a general rule covering the phenomenon, a rule that came to be called "Hubble's Law." This rule states that the speed of recession is proportional to the distance of a galaxy. If one galaxy is five times as far as another, the first is receding at five times the speed of the other.

Hubble's Law was based entirely on observation—on the measurement of red shifts. These observations had barely begun to be made, however, when a theoretical consideration of the matter was advanced.

In 1916, Einstein presented his General Theory of Relativity, which, for the first time, improved on Newton's view of gravitation. The theory included a set of "field equations" that could be used to describe the universe as a whole.

Einstein thought his field equations described a "static universe," one which, taken as a whole, was stable and underwent no change. In 1917, however, the Dutch astronomer Willem De Sitter (1872-1934) demonstrated that the equations could be interpreted to show that the universe was steadily expanding. This view of the "expanding universe" quickly grew more popular, and Einstein himself came around to the view.

The Big Bang

If the universe is indeed expanding, it is larger each day than it was the day before. If we imagine ourselves moving backward in time, however, as though we were running a motion picture in reverse, we can see that the universe must be smaller each day.

A universe can expand for an indefinite period forward in time so that there may never be any true end to it. A universe cannot contract for an indefinite period backward in time, however, for a contracting universe must eventually shrink to zero, and it can then contract no further. That zero must mark a beginning to the universe.

The first to make this clear was a Russian mathematician, Alexander Alexandrovich Friedmann (1888-1925), who advanced the notion in 1922, in the course of his mathematical analysis of the expanding universe. He died soon afterward, however, and could not follow it up.

Independently, however, the Belgian astronomer Georges Edouard Lemaître (1894-1966) advanced a similar notion in 1927. He supposed that, to begin with, all the matter in the universe was compressed into a tiny volume, which he called the "cosmic egg." This volume expanded violently and is still expanding.

When Hubble advanced his law in 1929 and described the observations on which it was based, it was clear that this was exactly what was to be expected of an expanding universe. To have all the galaxies receding from us—and doing so at a faster and faster rate, the further away they are from us—is no indication of anything special about us and our own galaxy. An expanding universe means that all the galaxies are receding from each other. If we were viewing the universe from *any* galaxy, not just our own, we would find that Hubble's Law would hold.

To be sure, the Andromeda galaxy and a few other nearby galaxies are approaching, but these are all part of the "Local Group." This is a cluster of galaxies that includes our own and Andromeda. These galaxies are bound to each other gravitationally and move about a common center of gravity, so that at any given time, some are approaching and some receding.

It came to be seen that the expanding universe does not mean that each individual galaxy recedes from all others but that each *cluster* of galaxies recedes from all other clusters. It is the clusters of galaxies that are the units out of which the universe is built.

The notion of the expanding cosmic egg was taken up and popularized by the Russian-American physicist George Gamow (1904-1968). He referred to the initial expansion as the *big bang*, an expression that caught on at once and is still used. It is the largest conceivable explosion that could take place in our universe, enormously larger than any mere supernova could be.

Gamow predicted that the radiation that accompanied the big bang should still be detectable as a low-intensity microwave radiation that could be noted in any direction—a radiation that should have certain calculable characteristics.

This suggestion was worked on further by the American physicist Robert Henry Dicke (1916-   ). In 1964, the German-American physicist Arno Allan Penzias (1933-   ) and a colleague, the American astronomer Robert Woodrow Wilson (1936-   ), detected this

143

The "big bang" is the most colossal explosion imaginable and may have created the Universe to about its present size in a fraction of a second.

"background microwave radiation" and found that it fit the theoretical predictions of Gamow and Dicke.

With this discovery, astronomers came to accept the existence of the big bang. It is now commonly supposed that the universe began as a very small object about fifteen billion years ago. The exact figure is still under dispute, but it can scarcely be less than ten billion years ago and may be as much as twenty billion years ago.

It seems to make more sense to suppose the universe

144

was created as a very small object that gradually evolved into the vast and variegated collection of clusters of galaxies that exists today, rather than to suppose that it was created, somehow, in the form in which it now exists. Nevertheless, there is still the question of how the universe was created in its original form as a very small object. Must we call upon the concept of a supernatural origin at this point?

Physicists are now working on the thought that the universe in its original tiny state may have formed out of nothing as a result of random process, even that there may be an infinite number of such tiny proto-universes continually being formed through the infinite volume of nothingness and that we live in one universe of countless many.

For the most part, however, physicists are content to trace the universe back to the big bang and to let it go at that. There is considerable uncertainty as to the initial stages of that enormous phenomenon, and as to how to go from the big bang to the universe as it now exists. The very early stages of universal evolution are still under dispute.

For instance, it was commonly supposed that the universe to begin with was infinitesimally small and at an infinitely high temperature, but that in unimaginably small fractions of a second, it grew large enough and cool enough to form the ultimate particles of matter—particles called *quarks*.

After another and longer period of time, say, $\frac{1}{10,000}$ of a second, the universe was large enough and cool enough for the quarks to come together in threes and form such subatomic particles as protons and neutrons. Then, after a still longer interval of several thousand years, the universe had cooled down sufficiently for protons and neutrons to combine with each other to form atomic nuclei, and for these to attract electrons and form intact atoms. After a yet longer interval of at least 100 million years, the stars and galaxies began to form and the modern universe (though still very small by present standards) came into being.

A modification of the big bang notion was advanced in the 1970s and is referred to as the "inflationary universe." Here, the original expansion is thought to have taken place very rapidly indeed, something that alters the details of the evolution of the universe in a number of ways.

One problem that arises is the fact that the universe is composed almost exclusively of normal matter made up of protons, neutrons, and electrons. It seems that these could not be formed without the simultaneous formation of their opposite numbers: *antiprotons*, *antineutrons*, and *antielectrons*. The latter group would combine to form *antimatter*, and it would seem that the universe should consist of equal quantities of both matter and antimatter—yet, as far as we can tell, it doesn't. It is almost entirely matter.

(And a good thing, too, for if the universe were made up of equal quantities of both matter and antimatter, the two would combine as fast as they were formed, annihilating each other and leaving only radiation behind. Our universe would not exist.)

New theories, referred to as "Grand Unified Theories" (or GUTS), have been worked out that account for the behavior of matter during the very high temperatures of the first instants after the big bang. They tend to show that there is a tiny asymmetry in the formation of matter. Ordinary matter is formed in an excess of a billionth over antimatter. When matter and antimatter combine and annihilate each other, that billionth part of matter is left over, and out of it the galaxies of the universe were formed.

Another great problem involving the big bang rests with the "lumpiness" of the universe. The big bang should have been spherically symmetrical; that is, the expansion should have been equal in all directions. In that case, the universe should consist of an evenly-spread-out mass of atoms, a kind of uniform gas. What made this gas clump together to form stars and galaxies?

The notion of the inflationary universe seems to offer an explanation for this lumpiness, and there may come a time when all the difficulties of the conception of a natural creation are ironed out.

7

THE ELEMENTS

Makeup of the Universe

It seems quite certain that, in the early period after the big bang, the tiny superhot universe expanded and cooled down sufficiently to allow protons and neutrons to combine with each other and form atomic nuclei. But *which* nuclei were formed and in what proportions? That is an interesting problem for cosmogonists (those scientists concerned with the origin of the universe), and it is one that will eventually lead us back to a consideration of novas and supernovas. Let us therefore consider it in some detail.

Atomic nuclei come in a number of varieties. One way of making sense out of those varieties is by classifying them according to the number of protons present in the nucleus. That number can be anywhere from one to over 100.

Each proton has an electric charge of +1. The only other particles found in nuclei are neutrons, which do not have an electric charge. The overall charge of the atomic nucleus is therefore equal to the number of protons it contains. A nucleus containing one proton has a charge of +1; one with two protons has a charge of +2; one with fifteen protons has a charge of +15; and so on. The number of protons in a given nucleus, or the number expressing the electric charge on the nucleus, is called its *atomic number*.

As the universe cools further, each nucleus is able to trap a certain number of electrons. Each electron has an electric charge of −1, and since opposite electric charges attract each other, a negatively charged electron tends to remain in the neighborhood of a positively charged nucleus. The number of electrons that can be held by an isolated nucleus under ordinary conditions is equal to the number of protons in the nucleus. With the number of protons in the nucleus equal to the number of electrons surrounding it, the overall electric charge of nucleus and surrounding electrons is zero, and the combination makes up a neutral atom. Both the number of protons and the number of electrons in such a neutral atom is equal to the atomic number of that atom.

Any substance that is made up of atoms that are all of identical atomic number is an *element*. For instance, hydrogen is an element, for it is made up only of atoms whose nuclei contain one proton and that include one electron in the neighborhood of that proton. Such an atom is a "hydrogen atom," and the nucleus of the atom is a "hydrogen nucleus." Finally, hydrogen has an atomic number of 1.

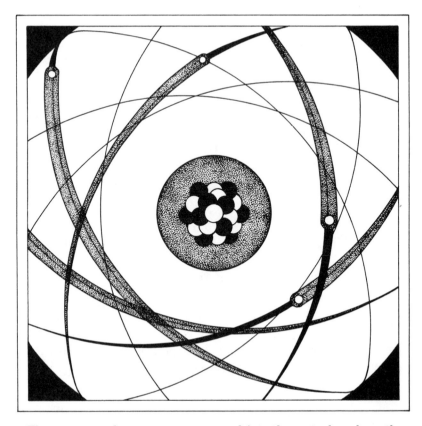

The protons and neutrons are squeezed into the central nucleus; the electrons are on the outskirts. This is a schematic picture; the true appearance cannot be drawn because it is like nothing with which we are acquainted.

Similarly, helium is an element that is made up of helium atoms that contain helium nuclei, each of which contains two protons, so that helium has an atomic number of 2. Similarly, lithium has an atomic number of 3, beryllium one of 4, boron one of 5, carbon one of 6, nitrogen one of 7, oxygen one of 8, and so on.

If we consider the material available to us for chemical analysis—the Earth's atmosphere, ocean, and soil—we find that there are eighty-one different elements that are stable; that is, eighty-one that will undergo no changes if left entirely to themselves during an indefinite period.

The least complicated atom on Earth (the least complicated that can exist, in fact) is that of hydrogen with its atomic number of 1. We can then climb the atomic numbers through 2, 3, 4 . . . until we reach the most complicated stable atom on Earth. That is the bismuth atom, which has an atomic number of 83, so that each bismuth nucleus contains eighty-three protons.

Since there are eighty-one stable elements, it is clear that in the list of atomic numbers from 1 (hydrogen) to 83 (bismuth), two numbers must be omitted, and that is true. Atoms containing forty-three protons or sixty-one protons are not stable, so that elements with atomic numbers 43 and 61 are not found in the natural materials that chemists have analyzed.

This does not mean that elements with atomic numbers of 43 and 61, or those with atomic numbers of more than 83, cannot exist *temporarily*. These atoms are not stable, so that they break down sooner or later, in one or more steps, to atoms that *are* stable. This does not necessarily happen instantly, and may even take a long time. Thorium (atomic number 90) and uranium (atomic number 92) take billions of years to undergo a substantial amount of breakdown to stable atoms of lead (atomic number 82).

In fact, during all the billions of years that Earth has existed, only part of the thorium and uranium that was originally present in its structure has had time to break down. Some 80 percent of the original thorium and 50 percent of the original uranium has so far escaped breakdown, and both can therefore be found in Earth's surface rocks today.

Though all eighty-one stable elements (plus thorium

and uranium) are found in substantial quantities in Earth's "crust" (its surface layers), they are not found in equal amounts by any means. The most common elements found there are oxygen (atomic number 8), silicon (atomic number 14), aluminum (atomic number 13), and iron (atomic number 26).

In terms of mass, oxygen makes up 46.6 percent of Earth's crust, silicon 27.7 percent, aluminum 8.13 percent, and iron 5.0 percent. The four together make up just about seven-eighths of the Earth's crust, while all the other elements together make up the remaining one-eighth.

These elements rarely exist in elemental form, to be sure. Atoms of different types are intermingled and tend to combine with each other, these combinations being called *compounds.* Silicon atoms and oxygen atoms bind together in complicated fashion, with atoms of iron, aluminum, and other elements clinging here and there to the silicon/oxygen combination. Such compounds are called *silicates,* and they make up the ordinary rocks that abound in the crust.

Since oxygen atoms are individually less massive than the other common elements of the Earth's crust, a given mass of oxygen contains more atoms than that same mass of the other elements would. Of every 1,000 atoms in the Earth's crust, 625 are oxygen, 212 are silicon, 65 are aluminum, and 19 are iron. Thus, 92 percent of the atoms in the Earth's crust are one or another of these four.

The Earth's crust is not a fair sample, however, of the universe, or even of the Earth as a whole.

The Earth's "core" (the central region, making up a third of the planet's mass) is thought to be largely iron. If that is taken into consideration, it is estimated that iron makes up about 38 percent of the mass of the entire Earth, oxygen makes up 28 percent, and silicon makes up

15 percent. The fourth most common element may be magnesium (atomic number 12), rather than aluminum, and that may make up 7 percent. The four together make up seven-eighths of the mass of the entire Earth.

If we go by atoms, then of every 1,000 atoms in the Earth as a whole, about 480 are oxygen, 215 are iron, 150 are silicon, and 80 are magnesium, so that these four, together, make up 92.5 percent of all the Earthly atoms.

Earth is not, however, a typical planet of the solar system. To be sure, Venus, Mercury, Mars, and the Moon are very similar to Earth in general composition, being made up of rocky materials and, in the case of Venus and Mercury, of an iron-rich core in addition. This may also be true, to some extent, of a few satellites and some of the asteroids, but all of these rocky worlds (with or without iron-rich cores) make up less than one-half of one percent of the total mass of all the objects that revolve about the Sun.

Fully 99.5 percent of the mass of the solar system (excluding the Sun) is to be found in the four giant planets: Jupiter, Saturn, Uranus, and Neptune. Of these, Jupiter alone, the largest, makes up a little over 70 percent of the total.

Jupiter may have a relatively small core that is rocky and metallic, but even if this is so, the bulk of the giant planet, judging from evidence gained from spectroscopy and planetary probes, is made up of hydrogen and helium. This would seem to be true of the other giant planets, too.

If we turn to the Sun, which is 500 times as massive as all the planetary bodies put together, from Jupiter down to the tiniest dust particle, we find (chiefly from spectroscopic evidence) that again hydrogen and helium make up its bulk. In fact, roughly 75 percent of its mass is hydrogen, 22 percent is helium, and 3 percent is all the remaining elements put together.

If we consider the makeup of the Sun in terms of the number of atoms, rather than mass, it would seem that of every 1,000 atoms in the Sun, there are 920 atoms of hydrogen and 80 atoms of helium. Less than one atom out of every thousand would represent all the remaining elements.

Since the Sun is so incomparably preponderant a portion of the solar system, we can't be far wrong in deciding that its elementary composition is representative of the solar system in general. The vast majority of stars resemble the Sun in elementary composition, and it has turned out that the thin gases that fill interstellar and intergalactic space are also largely hydrogen and helium.

Consequently, we are probably not wrong in judging that of every 1,000 atoms in the entire universe, 920 are hydrogen, 80 are helium, and less than one is everything else.

Hydrogen and Helium

Why is that? Does a hydrogen/helium universe fit in with the big bang?

Apparently, yes—at least according to a system of reasoning first advanced by Gamow, and since improved on, though not fundamentally changed.

Here is the way it works. Very soon after the big bang, a fraction of a second after, the expanding universe had cooled to the point where the familiar constituents of atoms were formed: protons, neutrons, and electrons. At the enormous temperature that still existed in the universe at that time, nothing more complicated could exist. The particles could not cling together; if they collided at that temperature, they simply bounced away again.

This remains true in the case of a proton-proton collision or a neutron-neutron collision even at much lower temperatures, such as that of the present universe. However, as the temperature of the early universe continued dropping, it finally became possible for a proton-neutron collision to result in a clinging together of those two particles. The two are held together by what is called the "strong interaction"—the strongest of the four interactions known to exist.

The single proton is a hydrogen nucleus, as I explained earlier in the chapter. The proton-neutron combination is, however, *also* a hydrogen nucleus because it contains one proton, and that is all that is required to qualify a nucleus as hydrogen. These two varieties of hydrogen nuclei—proton and proton-neutron—are referred to as "isotopes" of hydrogen, and are named according to the total number of particles each possesses. The proton, which is but one particle, is the nucleus of "hydrogen-1." The proton-neutron combination, which is two particles all together, is the nucleus of "hydrogen-2."

At the high temperatures of the early universe, when the various nuclei were being formed, the hydrogen-2 nucleus was not very stable. It tended to fall apart into individual protons and neutrons again or else to combine with additional particles to form more complex (but possibly more stable) nuclei.

A hydrogen-2 nucleus may collide with a proton and cling to it, forming a nucleus made up of two protons and a neutron. Since there are two protons present in this combination, it is a helium nucleus, and since there are three particles all together, it is "helium-3."

If hydrogen-2 collides with and clings to a neutron, a nucleus is formed consisting of one proton and two neutrons, again making three particles all together. The result is "hydrogen-3."

Hydrogen-3 is unstable at any temperature, even the cool temperatures of the present universe, so that it undergoes internal change even if it is kept from any interference by, or collision with, other particles. One of the two neutrons in the hydrogen-3 nucleus sooner or later turns into a proton, so that hydrogen-3 becomes helium-3. The change is not extraordinarily rapid under present conditions, half of the hydrogen-3 nuclei undergoing conversion to hydrogen-3 in a little over twelve years. At the enormous temperatures of the early universe, the change was undoubtedly more rapid.

Thus, we now have three types of nuclei that are stable under present-day conditions: hydrogen-1, hydrogen-2, and helium-3.

The particles in helium-3 cling together even more weakly than do those in hydrogen-2, and there is a strong tendency, at the elevated temperatures of the early universe, for helium-3 to break up or to undergo changes by further addition of particles.

If helium-3 were to encounter a proton, and if it were to cling, then we would have a nucleus made up of three protons and a neutron. This would be "lithium-4." Lithium-4 is, however, unstable at any temperature, and even at the cool temperatures of Earth's surface it undergoes a rapid conversion of one of its protons to a neutron. The result is a two-proton/two-neutron combination or "helium-4."

Helium-4 is a very stable nucleus, the most stable known at ordinary temperatures except for the single proton that makes up hydrogen-1. Once formed, it has little tendency to break up, even at very high temperatures.

If helium-3 collides with and clings to a neutron, helium-4 is formed at once. If two hydrogen-2 nuclei collide and cling, helium-4 is again formed. If a helium-3 collides with a hydrogen-2 or with another helium-3, then helium-4 is formed, while excess particles are split off as individual protons or neutrons. Thus, helium-4 is formed at the expense of hydrogen-2 and helium-3.

In essence, then, as the universe cooled to the point where protons and neutrons could combine to form more complicated nuclei, the first such nucleus that was formed in quantity was helium-4.

As the universe continued to expand and cool, however, hydrogen-2 and helium-3 grew less likely to change and some of each was, so to speak, frozen into continued existence. At the present time, only one hydrogen atom out of every 7,000 is hydrogen-2. Helium-3 is even rarer. Only one helium atom out of a million is helium-3.

We can ignore hydrogen-2 and helium-3 then, and say that soon after the universe had cooled sufficiently, it came to be formed of hydrogen-1 and helium-4 nuclei. Its mass was 75 percent hydrogen-1 and 25 percent helium-4.

Eventually, in places where the temperature was low enough, the nuclei would attract negatively charged electrons, which would be held to the positively charged nuclei by the "electromagnetic interaction," the second strongest of the four interactions. The one proton of the hydrogen-1 nucleus would be associated with one electron, and the two protons of the helium-4 nucleus would be associated with two electrons. In this way, hydrogen and helium atoms would be formed.

In terms of atom numbers, there would be 920 atoms of hydrogen-1 and 80 atoms of helium-4 out of every 1,000 atoms in the universe.

There you have the explanation of the hydrogen/helium universe.

But wait! What about the atoms more massive than those of helium and with a higher atomic number? (We can lump all the atoms containing more than four particles in their nuclei as "massive atoms.") There are very few massive atoms in the universe, but they do exist. How were they formed?

One apparently logical answer is that, although helium-4 is a very stable nucleus, there might be a slight tendency for it to combine with a proton, a neutron, a hydrogen-2, a helium-3, or another helium-4, to form a small quantity of various massive atoms and that this is the source of the 3 percent or so of the mass of the present-day universe that is made up of these atoms.

Unfortunately, this answer does not survive examination.

If helium-4 were to collide with hydrogen-1 (a single proton) and the two were to cling together, the result would be a nucleus of three protons and two neutrons. That would be "lithium-5." If helium-4 were to collide with, and cling to, a neutron, the result would be a nucleus of two protons and three neutrons or "helium-5."

Neither lithium-5 nor helium-5, if formed, would, even in today's cool universe, last for much more than a few trillionths of a trillionth of a second. In that interval, it would break up into helium-4 and either a proton or a neutron again.

The chance that helium-4 might collide with and cling to a hydrogen-2 or a helium-3 nucleus is remote, considering how sparsely these latter two nuclei are to be found in the primordial mix. Any massive atoms that might be formed in this way would be in far too small a quantity to account for those present today.

There is a somewhat better chance that a helium-4 nucleus might collide with and cling to a second helium-4 nucleus. Such a double helium-4 nucleus would consist of four protons and four neutrons and be "beryllium-8." Beryllium-8, however, is another exceedingly unstable nucleus and even in our present-day universe does not exist for more than a few hundredths of a trillionth of a second. Once formed, it falls apart into two helium-4 nuclei again.

To be sure, something useful might happen if *three* helium-4 nuclei meet in a three-way collision and cling to one another, but the chance of this happening in a mix in which the helium-4 is surrounded by a preponderance of hydrogen-1 is too small to consider.

Consequently, by the time the universe has expanded and cooled to the point where the formation of complex nuclei is over, only hydrogen-1 and helium-4 would exist in quantity. If any spare neutrons remain, they break down to protons (hydrogen-1) and electrons. *No massive atoms are formed.*

In such a universe, clouds of hydrogen-helium gas would break up into galaxy-sized masses, and these would condense into stars and giant planets. Stars and giant planets are almost all hydrogen and helium, after all. Is there any reason we ought to worry about massive atoms, then, which make up only 3 percent of the mass and less than 1 percent of the numbers of atoms in existence?

Yes! That 3 percent must be explained. Even if we ignore the small quantities of massive atoms in stars and giant planets, a planet like the Earth consists almost exclusively of massive atoms.

What's more, in the human body, and in living things generally, hydrogen is found to the extent of 10 percent of the mass only. No helium at all is present. The remaining 90 percent of the mass consists of massive atoms.

In other words, if the universe remained as it was when the process of nuclei formation in the aftermath of the big bang had been completed, planets like Earth, and life as we know it, would be completely impossible.

But for us to be here on this world of ours, massive atoms must have formed. How?

Escape from the Stars

Actually, this is no real puzzle for us, for we have already discussed, earlier in the book, the manner in which nuclei formation takes place at the core of stars. In our Sun, for instance, hydrogen is constantly being converted to helium in the central regions, and it is such hydrogen fusion that serves as the source of the Sun's energy. Hydrogen fusion is also proceeding in all the other stars on the main sequence.

If this were the only change going on, and if it continued to go on indefinitely at the present-day rate, then all the hydrogen would be fused and the universe would be pure helium in about 500 billion years (about thirty to forty times the present age of the universe). That, however, would still not account for the presence of massive atoms.

Massive atoms, we now know, are formed at the core of stars. But they are formed only when the time comes for such stars to leave the main sequence. By that climactic moment, the core has become so dense and so hot that helium-4 nuclei are smashing together with great speed and frequency. Every once in a while, *three* helium-4 nuclei would collide and cling together to form a stable nucleus made up of six protons and six neutrons. This is "carbon-12."

How can there be a triple collision at the core of stars now, but not in the time following the big bang?

Well, at the core of stars about to leave the main sequence, the temperature is approximately 100,000,000° C., and pressures are enormous. Such temperatures and pressures also existed in the very early universe, but there is one advantage the core has over the early universe: the core of main sequence stars is pure helium-4. It is much easier for a triple collision of helium-4 to take place when no other nuclei are present than (as in the case of the aftermath of the big bang) when most of the nuclei surrounding the individual helium-4 nuclei are hydrogen-1.

Thus, massive nuclei are formed at the core of stars all through the history of the universe, although such nuclei weren't formed immediately after the big bang. What's more, massive nuclei are still forming today at stellar cores and will continue to form there for many billions of years. Not only carbon nuclei have formed and will continue to be formed, but all the other massive nuclei up to and including iron—which, as I explained earlier, is a dead end for normal fusion processes in stars.

That leaves us with two questions:

1) How are the massive nuclei, after being formed at the center of stars, spread through the universe generally so that they end up being found on Earth and in our bodies?

2) How do elements with nuclei more massive than those of iron manage to be formed? After all, the most massive iron nucleus that is stable is iron-58, made up of 26 protons and 32 neutrons; yet there exist a number of still more massive nuclei on Earth, up to and including uranium-238, made up of 92 protons and 146 neutrons.

Let us tackle the first question first. Are there processes that act to spread stellar material out through the universe?

There are, and we can see some of them clearly when we study our own Sun.

To the unaided eye (using the proper precautions), the Sun might appear to be a quiet, featureless ball of light, but we now know it to be in a state of perpetual storm. The enormous temperatures in the deep interior set up convection movements in the upper layers (as in a pot of water on the stove that is approaching the boil). The solar substance is continually rising here and there and breaking the surface, so that the Sun's surface is covered with "granules," each one representing a convection column and each one about the area of a fairly large American state or European nation, though they look small on a photograph of the solar surface.

The convective material expands and cools as it rises, so that once on the surface, it tends to sink and be replaced by hotter material from below. The circulation never stops, and it helps deliver energy from the core to the surface. From the surface, energy is liberated into space in the form of radiation, a great deal of it as visible light; and, of course, life on Earth depends on that radiation.

The process of convection can sometimes lead to violent events on the surface, so that quantities of actual solar material, and not radiation only, may be hurled outward into space.

In 1842, a total eclipse of the Sun was visible from southern France and northern Italy. In those days, eclipses were not often studied in detail, for they usually took place in regions far removed from advanced astronomical observatories, and it was not easy to travel long distances with a full load of equipment. The 1842 eclipse, however, was near the astronomical centers of western Europe, and astronomers flocked to study it with their instruments.

For the first time, it was noted that around the rim of the Sun there were glowing, reddish objects that were clearly visible once the glare of the solar disk was obscured by the Moon. They looked like jets of material shooting out into space and were named *prominences*.

For a while, astronomers weren't certain whether the prominences emerged from the Sun or from the Moon, but in 1851, there was another European eclipse, one that was visible in Sweden. Close study made it plain that the prominences were solar phenomena and that the Moon had nothing to do with them.

Prominences have been studied attentively since then, and they can now be viewed with the proper instruments at any time so that there is no need to wait for a total eclipse. Some prominences arch upward mightily and reach heights of tens of thousands of kilometers above the solar surface. Some move upward explosively, at speeds of as much as 1,300 kilometers (800 miles) per second.

Though prominences are the most spectacular events one can follow at the Sun's surface, they are not the most energetic ones.

In 1859, the English astronomer Richard Christopher Carrington (1826-1875) noted a starlike point of light burst out upon the Sun's surface, last five minutes, and then subside. This is the first recorded observation of what we now call a *solar flare*. Carrington speculated that a large meteor had fallen into the Sun.

Carrington's observation did not attract much attention until the American astronomer George Ellery Hale (1868-1938) invented the spectrohelioscope in 1926. This made it possible to view the Sun in the light of a particular wavelength. Solar flares are noticeably rich in certain wavelengths of light, and when the Sun is viewed in those wavelengths, the flares show up brightly.

We now know that flares are quite common. They are associated with sunspots, and when the Sun is especially rich in such spots, small flares occur every few hours and major ones every few weeks.

Solar flares are energetic explosions on the Sun's surface, and those regions of the surface that are flaring are far hotter than the nonflaring portions that surround

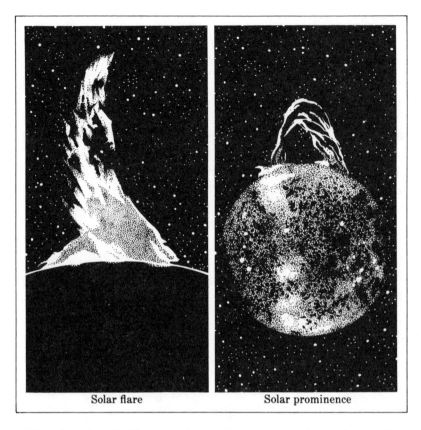

Solar flare | Solar prominence

Solar flares are the Sun's surface at the most energetic and can make themselves felt on Earth. Solar prominences are less energetic but more spectacular.

them. A flare that covers only a thousandth of the Sun's surface can send out more energetic radiation—such as ultraviolet light, x-rays, and even gamma rays—than the entire ordinary surface of the Sun would.

Though prominences look spectacular and can sometimes last for days, very little material is lost by the Sun through them. It is different with the flares, however. They are much less noticeable and many of them last only a few minutes; even the largest have totally subsided in a couple of hours. Nevertheless, so energetic are they that they shoot material out into space, material that is forever lost to the Sun.

This began to be understood in 1843, when a German astronomer, Samuel Heinrich Schwabe (1789-1875), who had studied the Sun almost daily for seventeen years, announced that the number of sunspots on the solar surface seemed to rise and fall in a period of about eleven years. In 1852, the British physicist Edward Sabine (1788-1883) observed that disturbances in the Earth's magnetic field ("magnetic storms") rose and fell in time with the sunspot cycle.

This was at first merely a statistical statement, for no one knew what the connection might be. In time, however, as the energetic nature of solar flares came to be understood, a connection was seen. Two days after a large solar flare had erupted near the center of the Sun's face (so that it was directly facing the Earth), the compass needles on Earth grew completely erratic, and the displays of the aurora became spectacular.

The two-day wait was important. If the effects on Earth were produced by radiation from the Sun, the time lapse between the flare and the effects would be eight minutes, for radiation travels from Sun to Earth at the speed of light. A delay of two days, however, means that whatever it is that produces the effect must travel from Sun to Earth at a speed of something like 900 kilometers (560 miles) per second. This is fast, but nowhere near the speed of light. It is the speed one would expect of subatomic particles; and if such particles carried electric charges and were hurled in our direction by events on the Sun, they would, as they passed Earth, produce the observed effects on the compass needles and on the auroras.

Once the notion of subatomic particles hurtling out of the Sun came to be grasped, there was a growth of understanding of another feature of the Sun.

When the Sun is totally eclipsed, one can see with the unaided eye a pearly radiance about the position of the Sun, with the black circle of the opaque Moon in the center. This is the Solar *corona*, from the Latin word for "crown," for the corona seems to surround the Sun like a halo.

In the same eclipse of 1842 that brought about the initial scientific studies of prominences, the corona was first studied attentively. It, too, was found to belong to the Sun rather than to the Moon. Beginning in 1860, photography was brought to the aid of coronal studies, and then the spectroscope was used for the purpose as well.

In 1870, during a solar eclipse in Spain, the American astronomer Charles Augustus Young (1834-1908) was the first to study the spectrum of the corona. He found a bright green line in the spectrum, one that didn't match the position of any known line for any known element. Other strange lines were also detected, and he assumed it represented an unknown element and called it "coronium."

Little could be done with "coronium," except to note the existence of the spectral line, until the nature of atomic structure was worked out. Every atom consists of a massive nucleus at the center, surrounded by one or more light electrons in the outskirts. Each time an electron is removed from an atom, the spectral lines produced by the atom change. Chemists could study the spectra of atoms from which a few electrons were removed, but the techniques for removing a great many electrons and studying the spectra under those conditions were not at first available.

In 1941, however, Bengt Edlen was able to show that "coronium" was not a new element at all. Common elements such as iron, nickel, and calcium produced lines like those in "coronium" once they had been stripped of a dozen electrons or so. "Coronium," therefore, represented ordinary elements with multiple electron deficiencies.

Such multiple deficiencies could only be brought about at very high temperatures, and Edlen suggested that the solar corona must be at a temperature of one or two million degrees. This was greeted with almost universal disbelief at first, but, eventually, when the days of rocketry dawned, it was found that the corona emitted x-rays, which it could only do if it were at the temperatures Edlen had proposed.

It would seem, then, that the corona is the outer atmosphere of the Sun, fed constantly by material hurled upward and outward by solar flares. The corona is very tenuous, containing less than a billion particles per cubic centimeter so that it is, on the average, not much more than a trillionth as dense as Earth's atmosphere at sea level. This makes it, actually, an extremely good vacuum. The energies hurled upward from the Sun's surface by flares, by magnetic fields, and by vast sound vibrations set up by the ever-rolling convection currents are divided among the relatively few particles. Although the total heat contained in the corona is small (considering its vast volume), the amount of heat crowded into each of the few particles is high, and it is this "heat per particle" that is measured as temperature.

The particles in the corona are the individual atoms that are hurled upward from the Sun's surface, with most or all of the electrons stripped away by high temperatures. Since the Sun is mostly hydrogen, most of the particles are hydrogen nuclei, or protons. Next to hydrogen in quantity are helium nuclei. All other more massive nuclei are very few in number. Even though some of the massive nuclei give rise to noticeable lines of "coronium," they are present only in trace amounts.

The particles in the corona move outward from the

Sun in all directions. As they move outward, the corona occupies a greater and greater volume and becomes more and more tenuous. Naturally, it delivers a fainter and fainter light in consequence, until, at some distance from the Sun, it can no longer be perceived.

Nevertheless, the fact that the corona fades out to unobservability doesn't stop it from continuing to exist in the form of outward-speeding particles. The American physicist Eugene Newman Parker (1927-   ) gave these speeding particles the name *solar wind* in 1959.

The solar wind extends past the inner planets. Rocket probes have even shown the solar wind to be detectable beyond the orbit of the planet Saturn, and it may well continue to be detectable beyond the orbits of Neptune and Pluto. All the planets, in other words, move about the Sun inside its vast atmosphere. However, so tenuous is that atmosphere that it doesn't interfere with planetary motion in any appreciable way.

The solar wind is not so tenuous, however, that it can't produce noticeable effects of other kinds. The particles of the solar wind are electrically charged, and it is these charged particles that are trapped in the Earth's magnetic field to form the "Van Allen belts," to produce auroras, and to affect Earth's compasses and electronic equipment. The solar flares momentarily strengthen the solar wind and, for a period of time, greatly intensify these effects.

In the neighborhood of the Earth, the particles of the solar wind are moving at speeds of 400-700 kilometers (250-435 miles) per second, and the number of particles varies from one to eighty per cubic centimeter. If these particles struck Earth's surface, they would have a most deleterious effect on life, but we are protected by Earth's magnetic field and by its atmosphere.

The amount of matter removed from the Sun by the solar wind is one billion kilograms (2.2 billion pounds) per

The charged particles trapped in Earth's magnetic field were unseen and unimagined until the dawning of the age of rocketry. (This drawing shows a cut-away view.)

second. This is a great deal by human standards, but it is the merest trifle to the Sun. The Sun has been on the main sequence for nearly five billion years and will continue to remain on it for five or six billion years more. If, through all that time, it loses mass to the solar wind at the present rate, the total loss over the Sun's entire lifetime as a main sequence star would be about 1/5,000 of its mass.

Nevertheless, 1/5,000 of the mass of a sizable star is no mean amount to be added to the general supply of matter that drifts through the vast spaces between the stars. This is the first example of how matter can be removed from stars and added to the supply of interstellar gas.

Nor is the Sun unusual in this respect. We have every reason to think that every star that has not yet collapsed emits a "stellar wind."

We can't study other stars the way we do the Sun, to be sure, but there are indications. There are, for instance, small, cool "red dwarfs," which, at irregular intervals, show sudden increases in brightness accompanied by a whitening of their light. The increase lasts from a few minutes to an hour or so and has all the properties one would expect of a flare on the small star's surface. These red dwarfs are therefore called *flare stars*.

A flare that is no larger than one on our Sun would have a much more noticeable effect on a small star. A flare that is large enough to increase the light of our Sun by 1 percent would be sufficient to increase the light of a dim star by 250 percent.

It may be, then, that red dwarfs may have stellar winds of respectable size.

Some stars appear to have unusually large stellar winds. Red giants have enormously extended structures, the largest having diameters about 500 times that of the Sun. This means that their surface gravities are comparatively low, for the greater mass of a large red giant is more than made up for by the unusually long distance from surface to center.

Besides this, red giants are approaching the end of their existence as extended stars and will soon collapse. They are therefore extraordinarily turbulent. One would expect strong forces to be driving material outward against the comparatively weak surface gravity.

The large red giant Betelgeuse is close enough to us for astronomers to be able to gather some details concerning it. Its stellar wind, for instance, is thought to be a billion times as intense as the Sun's. Even though Betelgeuse has sixteen times the mass of the Sun, its mass would, at that rate of disappearance, vanish entirely in a million years or so—except that long before then it would have collapsed.

We might suppose that, on the average, the solar wind of our own luminary is not too far removed from the average level of intensity of stellar winds in general. If we suppose that there are, perhaps, as many as 300 billion stars in our galaxy, the total mass lost by all the stellar winds could be some 300 billion billion (3×10^{20}) kilograms $(6.6 \times 10^{20}$ pounds) per second.

This means that every 200 years, an amount of matter equal to the mass of the Sun will have been transferred from stars to interstellar space. Assuming our galaxy to be fifteen billion years old and that stellar winds have been continuing at a steady pace through all this time, the total mass that has been transferred from stars to space equals the mass of about seventy-five million stars the size of our Sun, or 1/2,700 the mass of the galaxy.

However, the stellar winds are drawn from the surface layers of stars, and these surface layers are entirely (or almost entirely) hydrogen and helium. Therefore, the stellar winds consist entirely (or almost entirely) of hydrogen and helium nuclei and do not contribute massive nuclei to the galactic mix. The massive nuclei, formed in the stellar cores, remain there and are undisturbed by the formation of stellar winds far away on the stars' surfaces.

When a star has traces of heavy nuclei in the upper layers of its structure well outside its core (as our Sun does), the stellar wind will naturally include traces of these nuclei. However, such massive nuclei were *not* formed in the interiors of the stars in which they are

171

found, but were there at the time the star was formed. They were derived from some outside source—the source we are trying to find.

Escape by Catastrophe

If the stellar winds are not the mechanism by which massive nuclei are transferred from the stellar core to outer space, then we must look to the more violent events that take place after stars leave the main sequence.

Here we can promptly eliminate most stars. About 75 to 80 percent of the stars that exist are substantially smaller than the Sun. These remain on the main sequence anywhere from twenty to 200 billion years, depending on just how small they are. This means that none of the small stars that exist have yet left the main sequence. Even the oldest, those that were formed in the early days of the universe, during the first billion years after the big bang, have not yet had time to consume their hydrogen fuel to the point where they must leave the main sequence.

What's more, when a small star does leave the main sequence, it does so with a minimum of fuss. As far as we can tell, the smaller the star, the less violent the events that come about after leaving the main sequence. A small star will (as all stars eventually must) expand to a red giant, but in its case the expansion will produce a comparatively small red giant. Such a red giant will probably endure for considerably longer than more spectacular ones will, and it will eventually collapse, more or less quietly, into a white dwarf that will be less dense than those such as Sirius B.

The massive elements that form in the interior of a

small star, largely carbon, nitrogen, and oxygen, after remaining in its core throughout its life on the main sequence, will continue to remain within the white dwarf after the star's collapse. At no point will they be transferred in more than trace quantities to the reservoir of interstellar gas. Except in very special cases, then, any massive elements formed in small stars remain in those stars for indefinitely long periods of time.

Stars as massive as our Sun, give or take 10 or 20 percent, collapse to white dwarfs after only five to fifteen billion years on the main sequence. Our Sun, which will endure about ten billion years on the main sequence, is still on it because it was formed only about five billion years ago. Sun-like stars that are older than the Sun may well have left the main sequence by now, and all such stars that were formed in the infancy of the universe have certainly done so.

Stars in the mass-range of our Sun form larger red giants than small stars do, and those larger red giants, when they reach the point of collapse to white dwarfs, do so more violently than small stars do. The energy of collapse is likely to blast the outermost layers of the star into space and produce a planetary nebula of the type mentioned earlier in the book.

The expanding shell of gas formed by the collapse of a Sun-like star may contain up to 10 or 20 percent of the original mass of the star. However, the material is drawn from the outer reaches of the star, and even when such stars are on the point of collapse, the outer reaches are still essentially a mixture of hydrogen and helium.

Even if, as a result of the turbulence of a star on the point of collapse, massive nuclei are brought to the surface from the interior and are blasted into space as part of the shell of gas, the amount transferred into space is insufficient to account for more than a tiny fraction of the amount of massive nuclei that exist in interstellar gas clouds.

But while we are on the subject of white-dwarf formation, what about those special cases where white dwarfs are not a dead end and where they *can* serve as agents for the distribution of matter into space?

Earlier in the book, I discussed white dwarfs that are part of a close binary system and that can gain matter from a companion star that is expanding to the red-giant stage. Every once in a while, some of this matter undergoes fusion at the surface of the white dwarf, and the vast energies produced brighten it into what we see on Earth as a nova and blast the fused material out into space.

The material added to the white dwarf, however, is mostly hydrogen and helium from the outermost layers of the expanding red giant. The fusion converts the hydrogen into helium, and what is then blasted into space is a cloud of helium. In this case, too, if any massive nuclei beyond helium are received from the companion star or are formed in the fusion process, they are far too small in quantity to account for the amounts of massive nuclei in interstellar gas clouds.

Where does that leave us? Only one possible source of massive nuclei remains: the supernova.

The Type I supernova, as I described it earlier, arises out of a situation similar to that of ordinary novas. A white dwarf is receiving material from a close companion that is expanding into a red giant. The difference is that the white dwarf is near the Chandrasekhar limit of mass so that, eventually, the added mass shoves it beyond the limit. The white dwarf must collapse. In the process, massive fusion takes place within it, and it explodes.

Its entire structure, equal in mass to 1.4 times that of our Sun, tears itself apart and is converted into a cloud of expanding gas. We see it shine as a supernova for a while, but the radiation, intense though it is for a time, fades eventually. The cloud of gas remains behind, expanding for millions of years until it fades into the general background of interstellar gas.

The white dwarf explosion spreads a vast quantity of carbon, nitrogen, oxygen, and neon (the elements that are most common of all the massive nuclei) into space. In the course of the explosion, a certain amount of further fusion takes place so that small quantities of nuclei, which are still more massive than neon, also form.

Of course, very few white dwarfs are massive enough and close enough to a large companion star to produce a Type I supernova, but during the fourteen-billion-year life span of the galaxy, there have been enough such explosions to account for a substantial portion of the massive nuclei in interstellar gas.

The remaining massive nuclei exist in interstellar gas as a result of Type II supernovas. These involve, as I mentioned earlier, massive stars that are ten, twenty, even sixty times as massive as the Sun.

In the course of their existence as enormous red giants, these huge stars undergo fusion at the core to the point where iron nuclei are formed in quantity. This is the dead end beyond which fusion cannot go as an energy-producing device. At a certain point in iron production, therefore, the star collapses.

Even though the core of the star contains, in successively deeper layers, massive nuclei up to and including those of iron, the outer regions of the star still contain vast quantities of as yet untouched hydrogen, which was never under the conditions of high temperature and pressure that would have forced it into fusion.

When a giant star collapses, it does so very suddenly so that it undergoes a sudden and cataclysmic increase in

175

both temperature and pressure. All the hydrogen (and helium, too) that had hitherto lived a comparatively uneventful life, now undergoes fusion—and all at once. A vast nuclear explosion results, which we see as a Type II supernova.

The energy that is released can be and is pumped into nuclear reactions that produce nuclei more massive than iron. Such formation requires an energy input, but during the height of the supernova fury, this energy is available. Nuclei are formed, indeed, up to uranium and even heavier. There is enough energy available to form nuclei that are radioactive (that is, unstable) and that will eventually break down. In fact, all the massive nuclei that now exist in the universe are formed in a Type II supernova explosion.

To be sure, massive stars of the type that must necessarily produce a Type II supernova explosion are not common. Less than one star in a million, perhaps, is massive enough for the purpose. That, however, isn't as uneventful as it sounds. It still means that there are tens of thousands of stars in our galaxy that are potential Type II supernovas.

Since such giant stars can remain on the main sequence only a few million years at the most, we might wonder why they haven't all exploded and gone their ways long, long ago. The answer is that new stars are being formed all the time and some of them are very massive. The Type II supernovas that we now see are the explosions of stars that were formed only a few million years ago. The Type II supernovas that will be visible in the far future will be the explosions of large stars that do not yet exist today.

There may even be more dramatic supernovas. Until comparatively recently, astronomers were quite certain that stars with masses more than sixty times that of our Sun were not at all likely to exist. Still more massive stars, it was thought, would develop so much heat at their cores that they would blow themselves up at once despite their enormous gravities. This meant they could never form to begin with.

In the 1980s, however, it was realized that these considerations didn't take into account certain aspects of Einstein's theory of general relativity. Once those aspects were added to astronomical calculations, it began to appear that stars with diameters up to 100 times and masses up to 2,000 times that of the Sun might still be reasonably stable. What's more, there were some astronomical observations that made it appear that such super-massive stars actually existed.

Naturally, super-massive stars would eventually collapse and explode into supernovas that would produce much more energy over a much longer period than ordinary supernovas would. We might think of these super-explosions as "Type III supernovas."

A Soviet astronomer, V. P. Urtrobin, has looked back over astronomical records to see if he could find a supernova that would seem to be Type III in nature. He suggests that a supernova detected in 1961 in a galaxy in the constellation of Perseus is a case in point. Instead of reaching peak brightness in days or weeks, this supernova took a full year to reach its peak and then declined very slowly, still being visible nine years later. The total energy it produced was ten times that of an ordinary supernova. Even at the time, astronomers thought this unusual and were puzzled.

Such super-massive stars are extremely rare, but the quantity of massive nuclei they produce may well be a thousand or more times as great as those produced by ordinary supernova. That means they must contribute a substantial proportion of the massive nuclei in the interstellar gas clouds.

There may have been as many as 300 million supernovas of all kinds that have exploded in our galaxy during its lifetime (and a similar amount—allowing for differences in size, of course—in every other galaxy), and this is quite enough to account for the quantities of massive nuclei in interstellar gas, in the outermost layers of ordinary stars, and in any planets in addition to those of our own planetary system that may exist.

You see, then, that virtually all of Earth and almost all of ourselves consist of atoms that were formed inside stars other than our own Sun and that were distributed through space in earlier supernova explosions. We can't point to particular atoms and say which star formed them and exactly when it was exploded into space, but we know that they were formed in some far distant star and reached us through some explosion in the far distant past.

We and our world, then, are not only born of the stars but of exploding stars. We are born of the supernovas.

8

STARS AND
PLANETS

First-Generation Stars

The universe began in the big bang some fifteen billion
years ago. It began as an unimaginably tiny structure at
an inconceivably high temperature.

Very rapidly, it expanded and cooled. It consisted of
radiation (photons) and quarks, plus electrons and neutrinos, at first, but massive subatomic particles—such as
protons and neutrons—very quickly followed. As the universe expanded and cooled further, the protons and neutrons formed such nuclei as hydrogen-2, helium-3, and
helium-4, but nothing beyond that. After a few minutes,
the process was over and the universe's immense supply
of hydrogen and helium nuclei had been produced.

Further expansion and cooling for perhaps 700,000 years resulted in the temperature dropping to the point where negatively charged electrons could take up positions in the neighborhood of the positively charged protons and more complex nuclei, being held in place by electromagnetic forces.

In this way hydrogen and helium atoms were formed. Helium atoms remain single under all circumstances, but if two hydrogen atoms should collide at a low enough temperature, they would remain together forming a two-atom combination referred to as a "hydrogen molecule."

As the universe continued to expand and cool, the hydrogen and helium expanded with it in all directions. We might suppose that the universe would therefore consist of a uniform cloud of these mixed gases, steadily thinning out everywhere, since they would have to fill a larger and larger volume of space as the universe expanded.

For some reason, however, the cloud did not remain of uniform density; it did not remain homogeneous. Perhaps as a result of random fluctuations and consequent turbulence, the atoms drifted in such a way that there would be slowly rotating regions of greater than normal density separated by regions of less than normal density.

If the atoms could have continued to move randomly, this would have eventually evened out. The high-density regions would have lost atoms to the low-density regions so that there would have been a tendency toward renewed homogeneity. To be sure, random motion and turbulence would continue to form high-density regions, but their positions would fluctuate endlessly (like the high-pressure and low-pressure regions in our own atmosphere).

Once a high-density region forms, it can prove to be permanent. The intensity of the gravitational field in the high-density region increases as the density does. The gravitational field, as it grows more intense, overcomes the tendency of randomly moving atoms to spread out. Indeed, the high-density region could have an intense enough gravitational field to capture atoms from the low-density regions, so that the high-density regions become ever more dense and the low-density regions ever less dense.

In short, the even mixture of hydrogen and helium, over time, clusters into immense clouds of gas separated by near-vacuum.

These immense gas clouds have the mass and volume we associate with galaxies, or even with clusters of galaxies, and we might call them *protogalaxies*. Within these protogalaxies, there are further unevennesses developed through random movement of the atoms. Eventually, the protogalaxies consist of billions of smaller gas clouds separated from each other by virtually empty space. Just as the protogalaxies rotate relative to each other, the smaller clouds within the protogalaxies rotate relative to each other. (The rotations are in different directions, interestingly, and if all the rotations are added together, the different directions cancel out and the total rotation for the universe as a whole is zero.)

Each gas cloud has a gravitational field of its own. A gas cloud that is dense enough will have a gravitational field intense enough to exert enough of a pull on the cloud to cause it to begin to contract.

Once a gas cloud begins to contract, its density increases and the intensity of its gravitational field therefore also increases. The efficiency with which an intensifying gravitational field can bring about a contraction necessarily increases as well. In other words, once a gas cloud begins to contract, it must continue to contract more and more quickly.

As the cloud contracts, the pressures and temperatures at the center of the cloud increase. There comes a time when those pressures and temperatures are high enough to initiate nuclear fusion. The temperature of the cloud climbs rapidly until it becomes hot enough to radiate light. It is then no longer a gas cloud; it is a star.

Stars form all over the protogalaxies, and when the universe was about a billion years old, the protogalaxies of gas clouds had become galaxies of shining stars. Our own galaxy was one of them.

The galaxies, when they formed, were composed of hydrogen and helium only (mostly hydrogen). The stars that formed were also pure hydrogen/helium in structure and were "first-generation stars."

If we imagine all the gas clouds condensing into such first-generation stars, that would seem to end the process once and for all. First-generation stars are relatively small and quiet and can easily remain on the main sequence for fourteen billion years, so that they are still in existence today. Furthermore, even after they collapse, they will do so, relatively quietly, into white dwarfs.

There are, indeed, galaxies that seem to contain very little in the way of clouds of gas and dust and in which virtually all the stars seem to be first generation. In these galaxies, the distribution of gas clouds during the protogalaxy period may have been rather even, and the clouds themselves of relatively uniform size.

Second-Generation Stars

In some galaxies, however, including our own, the clouds of gas may have been, for some reason, uneven in size. The larger clouds would have condensed more quickly than the others, since the larger clouds would have more intense gravitational fields. Out of those larger clouds would form massive stars, which would be short-lived and which would explode as supernovas.

The supernovas would appear almost instantaneously on the long astronomic scale of time and would blast material out into space even while many gas clouds still existed that had not yet had time to condense into stars.

The energetic supernova material, mixing with the gas clouds, would heat them. The hotter a cloud, the more rapid are the random movements of the atoms making them up and, therefore, the greater the tendency for those atoms to move outward and dissipate. A cooling cloud, which is just beginning to condense under its own gravitational pull, would expand when heated in this way. Its gravitational field would grow less intense and the period when condensation would begin might be postponed for a long time, even indefinitely.

These early supernovas would, therefore, perform two functions. They would, first, keep gas clouds in being and prevent them from condensing, so that many galaxies even today are rich in such clouds. And they would, second, inject those clouds with nuclei more massive than helium. These massive nuclei can combine with hydrogen and with each other to form dust particles so that the clouds come to consist of gas and dust.

Thus, some galaxies, as they now exist, have no more than 2 percent of their total mass in the form of interstellar clouds of gas; others, where the supernovas have put in their work, may have as much as 25 percent of their total mass in the form of clouds of gas and dust.

In the cloud-rich galaxies, the clouds are not evenly distributed. Such galaxies are usually spiral galaxies, and the clouds of gas and dust are heavily concentrated in the spiral arms. It is in the spiral arms of our own galaxy that our Sun is located, and by some estimates, about half of the mass of these spiral arms of our galaxy is to be found in the form of interstellar clouds of gas and dust.

So dusty are the outskirts of the galaxy in which we live that we are seriously hampered in our view of its structure. In the plane of the Milky Way, where the clouds are concentrated, we can see nothing but the nearer stars. All else is blocked by the clouds. We cannot see the center of our galaxy by ordinary light, let alone any part of our galaxy beyond the center.

It is only because we have learned to make use of radio waves that can penetrate the clouds easily, and because the center of the galaxy is a highly active region that emits copious radio waves, that we have learned anything at all about that region.

The interstellar clouds that now exist in our galaxy have been exposed to the influence of millions of supernova explosions for fourteen billion years and are therefore appreciably stirred and enriched. About 1 percent of the atoms these great clouds contain (or 3 percent of the mass) consists of the massive atoms beyond helium that exist only because they arrived as part of the heavy atomic detritus driven outward into interstellar space by the incredible force of a supernova explosion.

Every once in a while one of these atomically enriched clouds of dust and gas, either in our own galaxy or in another, will begin to undergo contraction and form a new star, or several, or an entire cluster. Stars that form out of interstellar clouds that have an appreciable content of massive atoms are "second-generation stars"; that is, their structure is built up, to some small but measurable degree, out of the material that was brewed at the core of earlier stars that are now dead and gone—or at least exist no longer on the main sequence.

The Sun is a second-generation star, having been

formed 4.6 billion years ago, at a time when the galaxy was just about ten billion years old. It was formed out of a cloud that had been subjected to the incoming debris of supernova explosions for all those billions of years, so that the Sun contained appreciable quantities of massive atoms from the moment of formation, though it was still almost entirely hydrogen/helium in structure.

If a star like the Sun can form ten billion years after the big bang, there are stars that must have formed since. (This is undoubtedly so, since there are stars on the main sequence today that are so massive they can remain on the main sequence only a few million years, which means they must have formed no earlier than a few million years ago.) In fact, there must be stars that are forming now, even in our own galaxy, and even in our own neighborhood of the galaxy, so that we might expect to see evidence of that formation.

What about the Orion nebula, for instance? That cloud of dust and gas has a total mass of perhaps 300 times that of our Sun and there are stars in it, or the cloud would not glow as it does. The stars are hidden by the gas and dust that surround them, just as the frosted glass of an electric light bulb glows from the bright incandescent wire within but hides the wire so we cannot see it in detail. Still, indications are that the stars are very massive ones and must therefore be quite young. They undoubtedly formed out of the cloud, and there must be others forming now as well.

If such star formation is taking place, portions of the cloud are condensing, and as these regions contract, they become denser and more opaque. Light from the stars within the nebula, which penetrates other portions of the cloud and makes it glow, has difficulty passing through the condensing portions. There should, therefore, be portions of the Orion nebula, between ourselves and the stars of the interior, that show up as small, dark, more or less circular regions.

Such bits of dark circularity in the Orion nebula were pointed out by the Dutch-American astronomer Bart Jan Bok (1906-1983) in 1947. They are known as "Bok globules" as a result, and it is possible (though not certain) that they represent stars in the process of formation.

We might ask what causes interstellar clouds to begin condensing into stars when they have existed as clouds for billions of years *without* condensing. Perhaps random motions of atoms and dust within such clouds create a denser state that intensifies the gravitational field and starts the process, but it can't be a very likely situation or it would have happened billions of years before.

In fact, random motions might gradually dissipate a cloud and melt it into the near-vacuum of interstellar space. There is, after all, a very thin and attenuated system of gas and fine dust throughout interstellar space. This may represent, in part, material that was never picked up in the formation of stars and of interstellar clouds but also, in part, material that was dissipated outward from clouds.

The existence of such interstellar matter was first demonstrated by the German astronomer Johannes Franz Hartmann (1865-1936) in 1904. He was studying the spectrum of a particular star and found that its spectral lines were shifted, which was to be expected because the star was moving away from us. Hartmann found, however, that certain lines, representing the element calcium, were *not* shifted. The calcium, at least, was more or less at rest and therefore could not be part of the star.

Since there was nothing between the star and ourselves but "empty" space, the calcium had to be located in that space, which, it had to be concluded, was not entirely empty after all. The calcium had to be spread out extremely thinly, but as the light traveled from the star to ourselves across many light-years of space, it would pass an atom of calcium periodically and a photon of light would be absorbed. Eventually enough photons would be absorbed to give a noticeable dark line.

The Swiss-American astronomer Robert Julius Trumpler (1866–1956) showed, in 1930, that there was sufficient interstellar dust (incredibly thin though it might be) to dim distant objects appreciably.

We might conclude then that the interstellar gas clouds that still exist and maintain their identity after billions of years (such as the cloud from which our Sun formed, or the clouds that exist today) are in a delicate state of equilibrium. They are not dense enough or cold enough to begin the process of condensation, and not rare enough or warm enough to dissipate into the background of interstellar gas.

In order for a star to form out of such a gas cloud, then, something must take place that upsets the equilibrium, even if only slightly and temporarily. What can the unbalancing "trigger" be?

Several possibilities have been advanced by astronomers. In the Orion nebula, for instance, the large, hot young stars that are now in existence must have powerful stellar winds, to which our own solar wind is but the merest zephyr. As these stellar winds sweep out into the surrounding nebulosity, they push the dust and gas ahead of them and squeeze it together into a greater density than it would otherwise possess. This, in turn, intensifies the gravitational field in that portion of the cloud and begins the process of condensation, which further compresses the gas and dust, further intensifies the gravitational field, and so on. A Bok globule forms and, eventually, a star.

But then, how did those hot, young stars form? In

particular, how did the *first* star form in the Orion nebula before there existed powerful stellar winds within the nebula to initiate the process of compression?

There are several possibilities.

The interstellar clouds, like the stars themselves, are in motion and revolve majestically about the central regions of the galaxy, which contain most of the galaxy's mass. In its motion, an interstellar cloud may eventually pass near a massive, hot sun whose stellar wind sets off a first wave of compression and star formation.

Or else, two interstellar clouds may meet and push against each other slightly, initiating a tiny compression; or the two may simply merge to form an overlapping region of greater density than either separately. The gravitational field in the overlapping region intensifies and condensation begins.

It may even be that an interstellar cloud may pass through a region of space unusually distant from surrounding stars so that its temperature drops slightly. The atoms and particles it is composed of slow in their motion and drop toward each other in consequence, so that the cloud becomes denser and condensation begins.

These possibilities represent such gentle triggers, however, that it is doubtful whether star formation is at all likely at the rate it seems to be taking place. Can there be a more forceful trigger?

Yes! If a supernova explodes in the comparatively near vicinity of an interstellar cloud, the wave of matter exploding outward will smash into the cloud like a shock wave. It will be a much more violent event than anything taking place in the vicinity of an ordinary star, or than in the merging of two clouds, and there will be a fiercer compression of the cloud and a more certain beginning of the process of star formation.

To be sure, as I mentioned earlier in the chapter, a supernova explosion may heat an interstellar cloud and

*The colossal explosion of a supernova begins the process of star for-
mation.*

prevent it from condensing, but much depends on how
close the supernova is, how dense the cloud is to begin
with, and so on. Under some conditions, the heating effect of the supernova predominates and under others the
compressing effect; in the latter case, star formation is
effected.

Is it possible, then (we have no definite evidence and
it is only a possibility), that about 4.6 billion years ago a
supernova exploded only a few light-years, perhaps, from
an interstellar cloud that had, until then, remained in
equilibrium for ten billion years? Did the supernova pro-

duce enough compression to begin the process that ended with the formation of our Sun?

If this is so, we owe supernovas a triple helping of gratitude:

First, over the eons, supernovas filled space with the massive elements that would otherwise never have come into existence; elements essential to our world and ourselves and without which we would never have appeared (nor would, perhaps, any life anywhere else in the universe).

Second, the energy of supernova explosions kept vast numbers of interstellar clouds (including the cloud that eventually gave rise to our Sun) from condensing prematurely before the necessary infusion of massive elements could be gained.

Third, a nearby supernova explosion supplied the initial trigger that caused an interstellar cloud, now containing considerable quantities of massive elements, to condense into our Sun.

Formation of Planets

We can see how a star (or two, or even a cluster) might develop by simple compression of an originally diffuse interstellar cloud. But how does it come about that a single star, like our Sun, ends up surrounded by planets—bodies far too small to become stars?

There have been two classes of theories that have been advanced in explanation: 1) catastrophic and 2) evolutionary.

In catastrophic theories, stars are viewed as forming simply as stars—singly or with companion stars—and with no family of planets. Each star can (and almost always would) live out its full life on the main sequence, then expand into a red giant, and finally collapse. Throughout its existence, it would remain without planets.

At some time in the course of the star's life, however, a violent event would take place. A second star might approach and pass by. The enormous gravitational pull between the two stars would draw matter out of both and this matter would develop into a planetary family, perhaps for each of the two stars. Or else one star of a binary system might undergo a supernova explosion of a type that would leave only tiny fragments behind, and these tiny fragments would be caught by the companion star and become planets. In either case (or in other catastrophes that might be imagined), the planets are younger, perhaps even much younger, than the stars they circle.

Such catastrophes are bound to be very rare, and if the catastrophic theories of planet formation are true, then planets are a very uncommon phenomenon indeed. Our own solar system might be one of only a handful of such objects in the galaxy.

In evolutionary theories, the same process that forms the star also forms the planets. By such theories, the planets are just as old as the star they circle; and all the members of our solar system, for instance, from the central Sun to the farthest comet, would be the same age. By such theories, moreover, it would also follow that most, if not all, stars would have planetary systems.

Which of the two classes of theories is correct?

It is hard to say. No decision can be made based on actual observation. To this day we have not been able to study star formation at sufficiently close range to tell whether or not planets are being formed and, if so, just how. Nor can we, even yet, determine definitely whether planetary systems are very common (which means formation is evolutionary) or very rare (which means it is catastrophic). We can only argue for or against either type of theory by way of various theoretical considerations.

Arguing from theory, it turned out that both the catastrophic theories and the evolutionary theories as presented prior to the 1940s had serious deficiencies. So serious were the deficiencies, in fact, that thoughtful astronomers had to reject each. One might almost say that, so lacking were all the theories presented, the only reasonable conclusion one could come to with respect to the solar system was that it didn't exist.

In the 1940s, however, new versions of the evolutionary theory seemed to take care of the worst deficiencies, and a satisfactory scenario for the formation of the solar system was worked out. Let us concentrate, then, on the evolutionary view, the first versions of which, as described earlier, were brought to the fore by Kant and by Laplace, as the nebular hypothesis, in the mid-to-late 1700s.

The nebular hypothesis involved a property called "angular momentum." The interstellar cloud that condensed into the Sun was slowly rotating, to begin with, and angular momentum is the measure of the quantity of rotation. This quantity depends partly upon the speed of rotation and partly on the average distance of all parts of the object from the axis of rotation. According to a well-established law of physics, the total quantity of angular momentum in a closed system (one that isn't interacting with anything outside itself) must remain constant. As the interstellar cloud condensed, the average distance of all its parts from the axis of rotation decreased steadily. In order to balance this decrease and keep the total angular momentum constant, the speed of rotation had to increase.

As the condensing cloud increased its speed of rotation, the centrifugal effect caused its equator to bulge outward. Instead of being more or less spherical, as it was to begin with, the cloud became more and more pancake-shaped. Eventually, the equatorial bulge became extreme enough for a ring of matter to be thrown off—to be detached from the equator. This ring of matter condensed into a planet. The cloud continued to grow smaller and to rotate faster until another ring of matter was thrown off. The process was repeated until all the planets were formed. The rings of matter, as they condensed, were also rotating at an increasing speed and threw off still smaller rings that became satellites.

The nebular hypothesis sounded sensible and was popular through most of the 1800s. However, it was difficult to see how a ring of matter would condense into a planet instead of forming an asteroid belt or just diffusing into space. Worse yet, the various planets of the solar system contain 98 percent of all the angular momentum of the system; the Sun itself only 2 percent. Astronomers could not work out any way in which all that angular momentum could have been concentrated into the small rings of matter thrown off by the condensing cloud. As a result, the nebular hypothesis was virtually discarded, and, for fifty years, catastrophic theories (with their own difficult problems) grew popular.

In 1944, however, the German astronomer Carl Friedrich von Weizsäcker (1912-   ) worked out a modification of the nebular hypothesis. Instead of having the cloud rotate smoothly as a single body, he suggested that it rotated turbulently, forming a series of whirlpools. As the cloud condensed and became more and more pancake-shaped, the whirlpools grew larger and larger as they were located further and further from the center. Where neighboring whirlpools brushed against one another, the material in one collided with the material in the other, and individual bits of matter tended to coalesce. Bodies grew larger and larger at the places of coalescence and eventually planets formed, each planet's position about twice as far as the one next closer to the Sun.

Weizsäcker's theory made it quite easy to see how planets would form, eliminating the difficulty of trying to imagine how gaseous rings could coalesce into planets. But what about the queer distribution of the angular momentum of the solar system? Weizsäcker's theory was quickly refined through a consideration of the electromagnetic field of the Sun and the changes the field underwent as condensation took place. It then became possible to understand the transfer of angular momentum from the large central Sun to the small planets in the periphery of the solar system. Astronomers are quite confident, therefore, that they now have a grasp of the essential details of the formation of planetary systems.

Why do the individual planets differ so in size and in other properties?

If the Sun were a first-generation star, made up entirely of hydrogen and helium, the planets would be much more alike. The entire cloud would have been hydrogen/helium in composition and that would mean that the planets would be of similar composition to the Sun.

Helium and hydrogen (the former as single atoms, the latter as two-atom molecules) do not combine further and remain gases down to very low temperatures. The only thing that would hold them together would be gravitational forces.

Imagine a cloud of hydrogen/helium condensing. A continual tug-of-war between gravitational forces takes place that tends to hold the mass together and the random motion of the individual atoms and molecules that acts to spread the mass outward and to dissipate it. The larger the mass of condensing matter and the more it has condensed, the more intense the gravitation and the more tightly the body will hold together. The colder the mass, the slower the random motion of the individual atoms and molecules, the less the tendency to dissipate, and, again, the more tightly the body will hold together.

The Sun, as it formed, had no trouble holding together because it contains over 99 percent of all the mass of the solar system. Even though it is a ball of gas that would easily dissipate if conditions were right, and even after it underwent nuclear ignition and became very hot, so that the dissipative tendency was enormously strengthened, the Sun's extremely intense gravitational field held its structure together without trouble.

The planets, being built up of far smaller masses of hydrogen/helium, would have had greater difficulty in forming.

We can imagine the planets forming at varying distance from the developing Sun, some very close and some very far. All of them would grow slowly as their gravitational fields would only narrowly suffice to overcome the dissipative tendency. However, as the planets did grow, the increasingly intense gravitational field would more and more tend to overwhelm the dissipative tendency, so that the developing planet would begin to grow faster and faster (the "snowball effect").

Eventually, the planets would consist of sizable bodies of hydrogen/helium, which, as they condensed, would grow moderately hot at the center. Planets, however, wouldn't have nearly the temperature or experience nearly the pressure at their centers that the much larger Sun would experience at *its* center, so that no planet could undergo nuclear ignition and become a tiny star.

Nevertheless, the planets would be large enough to hold together even though the high temperatures in their depths would tend to increase the dissipative forces. Fortunately for the planets, their substance does not conduct heat well, so that even though the center is hot, the surface remains cold, and it is at the surface that the dissipative tendency can do its worst.

Perhaps the planets had largely completed the process of formation when the condensing Sun reached nuclear ignition and blazed out. When that happened, two new factors would be introduced.

First, the Sun would emit radiation that would heat the surface of the newly formed planets. Second, the Sun would emit a solar wind in all directions.

The warming of the planetary surface would increase the dissipative tendency where it would work most efficiently, so that clouds of hydrogen/helium vapor would rise from the planets. The solar wind would then sweep the vapor away from the planets.

Naturally, these two effects would be strongest near the Sun and would fall off with distance. The planets that had formed nearest the Sun would have the greatest tendency to vaporize and would be strongly subject to the sweeping-away effect of the solar wind. Those nearby planets would decrease in mass, therefore. As they did so, their gravitational fields would decrease in intensity and both the vaporization and the sweeping away would accelerate. Finally, the nearby planets would be completely dissipated.

At greater distances from the Sun, the heating and sweeping-away effects would be minor and the relatively massive planets would survive. Satellites of those planets, had they formed, might not survive due to their much weaker gravitational fields.

We can conclude, then, that if the Sun were a first-generation star, it would have had a few planets corresponding in distance and in general chemical composition to the gas-giants, Jupiter, Saturn, Uranus, and Neptune—but nothing else. There would be no planets on which human beings could conceivably live and no material out of which living tissue could be formed. A planetary system circling a first-generation star would be absolutely without life as we know it.

Formation of Earth

The Sun is a second-generation star, thanks to the existence of supernovas. That means that the interstellar cloud out of which the solar system actually formed was built up of four kinds of substances.

First, there is hydrogen and helium, which made up 97 percent of the mass of the original cloud—even though it is a second-generation one.

Second, there are those massive elements that are only a little more massive than hydrogen and helium—carbon, nitrogen, and oxygen being the most common. These combine with hydrogen to form methane, ammonia, and water, respectively. Of the three, water freezes most readily, forming ice. As the temperature drops further, first ammonia and then methane freeze to form substances that resemble ice very much in appearance. At the low temperature in which the planets first took shape, all three of these compounds (together with other similar, but less common, ones) probably existed in the frozen state, and they are referred to, generally, as the *ices*.

Third, there are still heavier elements, such as aluminum, magnesium, silicon, iron, and nickel. Aluminum, magnesium, and silicon (together with other less common elements) can combine with oxygen to form "silicates." It is the silicates that make up the rocky portions of the Earth.

197

Fourth, iron and nickel atoms can also participate in the formation of silicates but are frequently plentiful enough to come together in relatively pure form with lesser quantities of some other similar substances. They can be referred to as "metals."

It might seem that with the original cloud 97 percent hydrogen/helium in mass, the small quantity of massive elements present would scarcely suffice to form a planet such as Earth, so that we would be no better off with a second-generation star than with a first-generation one. However, the total mass of the solar system is equal to 343,600 times that of Earth. If 3 percent of that total are massive elements, then there are enough such elements available to build more than 10,000 planets such as Earth.

To be sure, over 99 percent of the massive elements remain in the Sun, but all the planetary material circling the Sun taken together is 448 times the mass of the Earth. If 3 percent of that are massive elements, there are still enough such elements available to build over thirteen planets the size of the Earth.

There is, quite simply, enough, and a planet like Earth can therefore form about a second-generation star like the Sun.

When planets of a second-generation star are forming, rock and metal will coalesce first. Silicate molecules and metal atoms cling together tightly, thanks to electromagnetic forces among their electrons, so that they do not depend on gravitation to hold them together. They will even cling together in small masses at very high temperatures of up to two or three thousand degrees.

Every planet, therefore, is likely to have a rock/metal core. At first the rock and metal are intermixed, but as the planet increases in size and grows hotter at the center, it becomes easier for the two to separate—especially if it gets hot enough for the metal to melt. Rocks have higher melting points than metals, to be sure, but even though the rocks may not melt, they will grow hot enough to be comparatively soft. Metal, since it is denser than rock, slowly drifts downward. Metal will therefore collect at the planetary center while the rock substances will remain as an envelope about the metal.

Thus, Earth has a metal core with rock enveloping it, and so have the planets Venus and Mercury. Mars and the Moon collected relatively little metal for some reason we cannot, as yet, explain. What metal exists remains mixed with the silicates, so that these two latter worlds stay essentially rocky through and through.

Once a metal/rock core has formed, it is that much easier, thanks to the gravitational field of that core, for the developing planet to collect a layer of ices about itself, and then a layer of hydrogen/helium about the ices. It is quite likely, then, that planets form more quickly about second-generation stars than about first-generation ones.

What happens, then, when the Sun ignites? The planets nearer to the Sun heat up on their surfaces and are faced with the buffeting of the solar wind.

Any hydrogen/helium that the inner planets collected, together with most or all of the ices, would be vaporized and swept away. The metal/rock cores, however, would cling together despite heat and solar wind.

Mercury would be so hot, and the Moon so small, that everything on their surfaces would be swept away. The same is true of the asteroids, which may have been larger and fewer at the time of the Sun's ignition. Venus and Earth were large enough, and Mars far enough away from the Sun, to hold on to a minor part of the ices, perhaps in loose combination with the silicates at first. All were able to retain substances that now form atmospheres. Earth is larger than Mars and cooler than Venus, so it could keep enough water to form the oceans.

Beyond the asteroids, the planets were not appreciably affected by the Sun's radiation and solar wind, and they retained most or all of the ices and hydrogen/helium envelope they had accumulated. So we have Jupiter, Saturn, Uranus, and Neptune; except for containing minor quantities of the massive elements, they are just what they would have been even if they had been formed in the neighborhood of a first-generation star and were circling it.

It is also possible for smaller bodies to form in the coolness and safety of the outer solar system. Some are largely rocky, such as Io, the innermost large satellite of Jupiter. Some are largely icy, such as Ganymede and Callisto, two other satellites of Jupiter; Titan, a satellite of Saturn; and very distant bodies, such as Pluto and the comets. Some are made up of both rock and ice, as is Europa, the fourth satellite of Jupiter.

In any case, Earth formed at just the right place and with just the right composition to make possible the formation of life—something that would simply not have been possible without the existence of supernovas.

9

LIFE AND
EVOLUTION

Fossils

Our profound debt to supernovas does not begin and end
with the formation of the Earth. We must also consider
the role that supernovas play in the formation and development of life, and in order to do that, we must now
change our focus from astronomy to geology and biology.
Let us begin by considering our planet's past.

Efforts were made, in the course of the past two centuries, to determine the age of the Earth, but it was not
until the discovery of radioactivity in 1896 that the
chance of making more than reasonable guesses was
granted geologists.

In 1907, the American chemist Bertram Borden Boltwood (1870–1927) suggested that since uranium broke down to lead at a steady and very slow, but easily calculated, rate, it should be possible to calculate the length of time a particular rock had remained solid and undisturbed by determining the quantity of uranium and lead in it.

Methods were indeed evolved to measure age by the breakdown of uranium and by other slow radioactive changes. By making such measurements, it was finally determined that the age of the solar system, and of the Earth in particular, was 4.6 billion years. At least that is how long ago the original cloud of gas and dust condensed into sizable solid objects that are still in existence.

Actually, since Earth has undergone all kinds of geological changes in the course of its lifetime, it is unlikely or, perhaps, quite impossible to locate rocks that have remained undisturbed from the very beginning of the planet's existence. The oldest Earth rocks yet found are about 3.4 billion years old, so that we have no direct record of the Earth's first billion years.

The Moon, smaller than Earth and less "alive" geologically speaking, has yielded rocks that are as much as 4.4 billion years old. Even the Moon, however, has not been left entirely undisturbed from the beginning. In the first few hundred million years of existence, both Earth and Moon were heavily bombarded by smaller bodies, as the process of formation of the two worlds was completed. The marks of the bombardment no longer exist on Earth, thanks to the action of wind, water, and life, but on the Moon they have left behind the clear evidence of numerous craters that mark collision sites.

Fortunately, meteorites are small bodies that have been undisturbed almost from the beginning, and it is the analysis of them that best justifies the 4.6-billion-year age of the solar system.

Life is not a very recent phenomenon on Earth. Life has been found on Earth through much of its long history, as has been made directly evident by the finding of fossils in rocks. These fossils are petrified remnants of portions of ancient life forms, and since they are incorporated in rock strata well below the surface, they are therefore presumed to be ancient.

Such fossils were reported even in ancient times, but through much of Western history they were ignored, or explained away in various implausible fashions, because for a while the dominant system of thought would have it, quite dogmatically, that the Earth, and the whole universe, was only a few thousand years old. Even scientists were reluctant to abandon, or contradict, that dogma.

In the course of the 1800s, however, scientists were forced to accept the fact that the Earth was very old.

Even though scientists could not yet determine the absolute age of the fossils, they could determine the relative age. They could tell which rocks were older by determining how deep below the surface the layer (or "stratum") in which a particular rock was found. It seemed a natural assumption to suppose that layers of sediment gradually, and very slowly, covered the Earth's surface as time went on, so that the deeper below the surface a particular rock stratum lay, the older the rocks in that stratum.

Once the relative age of the strata was determined, the relative age of fossils could be determined by noting in which stratum each fossil was found.

The oldest fossil-bearing rocks were those given the name "Cambrian" by the English geologist Adam Sedgwick (1785-1873). He named it in honor of "Cambria," the ancient Roman name for the region we now call Wales, since Sedgwick had first studied rocks of this type in Wales.

The Cambrian fossils, it was quite clear, were the remains of sea organisms. No signs of land life existed in the fossil record of that period. The dominant form of life was various types of a kind of shellfish, and these were given the name of *trilobites*. The living animal that most closely resembles the ancient trilobite is the horseshoe crab.

All rocks more ancient than the Cambrian are lumped together as "Pre-Cambrian."

With the development of age determinations through radioactive breakdowns, it became clear that the

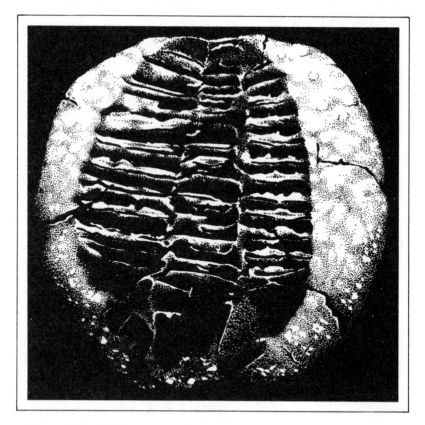

The mark left behind by a sea creature that died half a billion years ago.

oldest Cambrian rocks and, therefore, the oldest fossils, were 600 million years old. This is a tremendous age, but it appeared that even the oldest fossils were comparatively recent when judged against the total age of the Earth.

In the rocks laid down during the first four billion years of Earth's history (seven-eighths of the Earth's lifetime), no fossils are to be found. Could it be that life existed on Earth only during the most recent eighth of its existence?

Biologists could not believe that. Fossil formation is a very chancy thing and takes place only under unusual circumstances. Uncounted billions of organisms have lived and died without leaving anything behind that was petrified and preserved in fossil form. It might well be that, by chance, whole groups of organisms may have left nothing behind that has survived to the present day to be found. On the other hand, some less common organisms may, by chance, have left rich hordes of fossil remnants.

Then, too, some parts of organisms are more easily fossilized than others. Teeth, bones, and shells, the "hard parts" of an organism, are far more easily fossilized than the soft tissues. Thus, from 50,000 to four million years ago, there were human-like organisms that roamed Africa and Eurasia, but we have very few fossil remmants that represent them—they were too intelligent to be often surprised by death under fossil-forming conditions—and those remnants that do exist are, for the most part, petrified hard parts, particularly skulls and teeth.

Among the earliest fossils, the trilobites are shelled organisms and are already quite complicated in structure.

In general, the more ancient an organism, the less advanced it is and the less complicated in structure. It is a natural assumption that prior to the Cambrian era, there will still more ancient forms of life that were less advanced than the trilobites; sufficiently less advanced to have no hard parts; to be as soft, through and through, as modern earthworms or slugs. They would have left virtually no fossil remains, so that the absence of fossils need not necessarily indicate "no life" but merely "no hard parts."

In the 1950s, the American biologist Elso Sterrenberg Barghoorn (1915-1984) found traces of fossilized colonies of blue-green algae near Lake Superior. Blue-green algae are among the simplest forms of cellular life that now exist. They are very much like bacteria, except for the fact that the blue-green algae contain chlorophyll, which bacteria do not.

Both blue-green algae and bacteria are made up of particularly small cells that do not have a distinctly separate nucleus but have nuclear material scattered through the cell generally. They are called *prokaryotes*, from Greek words meaning "before the nucleus." All other cells, from one-celled plants and animals to those cells out of which multicellular organisms (including ourselves) are formed, are *eukaryotes*, from Greek words meaning "true nucleus."

Fossil blue-green algae are not easy to spot. They are so small that they must be studied under the microscope, and the tiny cells must be identified as such by delicate structural signs that can be shown to be of biological, rather than mineralogical, origin.

It was not an easy task, but Barghoorn was able to present the evidence in meticulous, and ultimately convincing, fashion. The first "microfossils" he located and studied were in rocks that were two billion years old. Once he knew what he was looking for, he found evidence of very simple microscopic life in older and older rocks. In 1977, he found such microfossils in South African rocks whose age was measured at 3.4 billion years.

Formation of Life

It would seem, then, that Earth was formed 4.6 billion years ago, but, for the first few hundred million years of its existence, it was kept in constant turmoil, due to the steady bombardment of its surface by the sizable fragments of matter that still circled the Sun in Earth's orbit and that periodically collided with both the Earth and Moon.

About four billion years ago, Earth was sufficiently at rest, and sufficiently in its present form, to be a habitable world. Within no more than half a billion years thereafter, it would appear, simple life had come into existence. For the remaining 3.5 billion years (three-fourths of its total existence) Earth has been, as far as we can tell, continuously inhabited by a variety of living organisms.

How was life first formed?

The only possible scientific conclusion (one that does not involve supernatural action, of which we have no evidence whatsoever) is that random combinations of simple molecules that existed in Earth's atmosphere and oceans built up more complicated and still more complicated molecules. Eventually, molecules formed that were sufficiently complicated to possess the properties we associate with life.

This is not something we can observe directly, either here on Earth, where we are separated from the event by billions of years of time, or on other worlds, since the nearest other habitable world must be separated from us by many light-years of space. Nevertheless, we can obtain indirect evidence.

To begin with, we must determine what the simple molecules that existed on the primordial Earth must have been. That they were the molecules that make up the ices, scientists now generally agree. There is some dispute on the exact combination, though. Water was certainly present, as were molecules containing nitrogen and others containing carbon.

On Jupiter and other worlds of the outer solar system, carbon and nitrogen are each present in combination with hydrogen—methane and ammonia, respectively. On Venus and Mars, carbon is present in combination with oxygen (carbon dioxide), while nitrogen atoms exist in pairs as nitrogen molecules.

There are some scientists who think the primordial atmosphere of Earth was ammonia, methane, and water vapor, with ammonia dissolving in large quantities in the ocean. Others think the primordial atmosphere of Earth was carbon dioxide, nitrogen, and water vapor, with carbon dioxide dissolving in substantial quantities in the ocean. It is also possible that the atmosphere was ammonia, methane, and water vapor (Atmosphere I) at the outset, this being converted by natural processes, not involving life, to carbon dioxide, nitrogen, and water vapor (Atmosphere II).

The choice between atmospheres is not a crucial one. In either atmosphere, the atoms of hydrogen, carbon, nitrogen, and oxygen (which make up 99 percent of the atoms of soft tissue in any organism) are to be found. The atoms that make up the remainder of the tissues, including the atoms that make the hard tissues hard, were to be found in solution in the primordial ocean.

Given the simple molecules (whichever they are), what processes would be required to build up more complicated molecules from them? Simple collisions and random interchange of atoms would not be enough. In general, the conversion of simple molecules into more complicated ones is an energy-consuming change. In other words, energy would have to be supplied to the system to make the change possible.

The primordial Earth, however, had numerous sources of energy available. There was the heat of volcanic action, or the electrical energy of the lightning bolt, and it is quite probable that, at the beginning, the Earth was a more violent place than it is now, with more volcanic eruptions and many more thunderstorms.

Then, too, there was the energy of radioactivity, and, at the start, radioactive intensities were greater than now because, in the billions of years that have passed since the Earth was formed, an appreciable fraction of the original supply of radioactive atoms has broken down.

Finally, there was the Sun's ultraviolet light. Nowadays, little of the ultraviolet light from the Sun reaches the Earth's surface because the oxygen of the atmosphere (consisting of molecules made up of two oxygen atoms each) is converted to ozone (which consists of molecules made up of *three* oxygen atoms each) high in the atmosphere. This ozone layer, some twenty-five kilometers (fifteen miles) high, is opaque to most of the ultraviolet so that little of it reaches the Earth's surface.

Oxygen, however, is not a natural constituent of the atmosphere. It is too active and combines with many other substances, so it would soon disappear from the atmosphere if it were left to itself. The only reason it does not disappear from the atmosphere is that green plants are constantly forming oxygen. Such plants use the energy of sunlight to combine carbon dioxide and water to form starches and other substances that the animal world can use as food. Oxygen is produced and discharged into the atmosphere as a by-product of the process.

In the primordial Earth, before life existed, there were no green plants and no oxygen-forming process. Hence, there was no oxygen in the atmosphere and no

The ozone layer is thin and rarefied but performs a vital function in protecting life from cosmic radiation. (This drawing is not in scale.)

ozone in the upper atmosphere. This means that ultraviolet light from the Sun could freely penetrate to Earth's surface.

In 1952, the American chemist Stanley Lloyd Miller (1930–   ) began with carefully purified and sterilized water and added an "atmosphere" of hydrogen, ammonia, and methane, thus duplicating a variety of Atmosphere I. He circulated this through his apparatus past an electric discharge, which represented an energy input that would mimic the effect of lightning. He kept this up for a week, then separated the components of his water solution. He found that simple organic compounds had been formed, including a few "amino acids," which are the building blocks of proteins that are, in turn, among the key components of living tissue.

Others repeated the experiment with ultraviolet light as the source of energy, and they obtained much the same results. Still others used varieties of Atmosphere II and also formed more complicated substances.

The Sri Lanka-born American biochemist Cyril Ponnamperuma (1923-   ) has been the most assiduous in working with this type of experiment. He has carried through the formation of "nucleotides" from simple compounds, these nucleotides being the building blocks of "nucleic acids," which are the other key component of living tissues. He has also formed adenosine triphosphate (ATP), which is a key substance as far as energy-handling in living tissue is concerned.

All the compounds that formed abiogenetically (without the intervention of life—except for the experimenter himself, of course) from samples of what might be the primordial atmosphere seem to be in the direction of living tissue.

The American biochemist Sidney Walter Fox (1912-   ), working in a different direction, began with a mixture of amino acids and, subjecting them to heat, formed protein-like substances. These, dissolved in water, formed tiny spheres that shared some properties with cells.

Experiments have not gone very far or come anywhere near a system that might be considered as living, in even the most primitive fashion. In the laboratory, however, work is done with small quantities of fluid over small periods of time, and yet surprising advances (admittedly small) *have* been made in the direction of life.

What if we imagined a whole ocean of simple compounds subjected to energy for hundreds of millions of years? It is, in that case, not hard to imagine a period of "chemical evolution" that ended at last with primitive living cells no longer than 3.5 billion years ago.

Development of Species

How many different times did life have to be formed? Were blue-green algae formed out of one pathway of chemical evolution and bacteria out of another? Did each kind of blue-green algae and bacteria form out of a totally separate pathway? Was there still another, more complicated set of pathways of chemical evolution that ended each in a different kind of trilobite? In a different kind of dinosaur? In a human being?

That seems utterly unlikely. If there were millions of different pathways of chemical evolution, one for each type of plant or animal or microorganism, even those that came into being quite recently, then there should be compounds undergoing chemical evolution right now. There are no such signs at all.

Besides, while one can understand chemical evolution proceeding in a world with a primordial atmosphere and no life, it is illogical to suppose it to be proceeding in an oxygen atmosphere and in a world containing life. Oxygen is an active substance that would combine with compounds complicated enough to be approaching life, break them up and destroy them. (Such compounds in living organisms today must be protected from oxygen in a variety of intricate fashions.) Then, too, once life came to exist, any compound that evolved into near-life would be suitable as food for some creature and would be promptly eaten.

Consequently, it makes considerable sense to suppose that life was formed only once in primordial times—or possibly several times, with all but one of the attempts not quite persisting. Once a particular life form had come into existence, persisted, and flourished, that very likely put an end to chemical evolution.

If so, why was not that one life form the only life form in existence from the time of its origin to the present day? How did it come about that there were so many different forms of life in the past (judging from the fossil evidence) and so many different forms of life in the present?

If the fossils are studied, it can be seen that there are apparent relationships, to a greater or lesser degree, between different life forms. Ancient organisms seem similar to certain modern organisms in certain ways, and between the two are often a series of other fossils that indicate organisms that have undergone changes that seem to lead from the ancient to the modern. This is supported by a variety of other types of evidence, biochemical and even observational.

The answer is that, little by little, as organisms reproduce themselves—parents giving birth to young, who grow up and in turn give birth to young—they change. Some kinds (or "species") become extinct. Some alter, little by little, into an organism sufficiently different to be considered another species. Some kinds give rise to two different descendant-species, or to more than two. The result is that the two million or so species of organisms that are estimated to be in existence today (including the one species of human beings, *Homo sapiens*) are the descendants of earlier species that are, in turn, the descendants of still earlier species, and so on back to the simple forms of life 3.5 billion years ago or so, and, through them, to the initial form of established life that resulted from the still earlier period of chemical evolution. The slow development of life from the simplest original form to the vast multitude of species, alive and extinct, is referred to as "biological evolution."

It was hard for earlier scientists to accept the notion of biological evolution for two reasons.

First, the prevailing religion of the Western world insisted on the literal words of the Bible, which not only seemed to show that the Earth was formed only a few thousand years ago but that each species was especially created by supernatural action, so that all species existed and were distinct from the start. To espouse biological evolution would, it was thought, shake the foundations of religion, and most scientists were sincerely religious and did not want to upset those foundations. Even were they iconoclastic enough to prefer careful reasoning to blind faith, earlier scientists might well have feared the response of an angry society.

Second, even where scientists were convinced that evolution must have taken place, there seems to have been no mechanism by which it could. Cats gave birth to kittens, dogs to puppies, human beings to babies, but there seemed to be no sign of any distinct change with the generations that would point to continuing evolution.

The first scientist to propose a mechanism was a French naturalist, Jean Baptiste de Lamarck (1744-1829). In 1809, he suggested that organisms overused some parts of their bodies and underused others. The overused parts developed and the underused parts dwindled, and the development and dwindling were transmitted to their offspring, who might continue the process, passing the results on to their offspring, and so on.

Thus, a particular antelope, feeding on leaves, would be continually stretching to reach leaves higher up. After years of this, his neck would have grown slightly longer, through stretching, and so would his legs. He would transmit his longer neck and legs to his offspring, who would continue to stretch. Eventually, after many generations, the antelope would have become a giraffe. So many generations would be required that the change would not be noticeable in a human lifetime or even during the period of human history.

This suggestion—evolution by the inheritance of acquired characteristics—was, however, wrong.

In the first place, acquired characteristics were not inherited, as was indicated by actual experiment. In the 1880s, as one example of such an experiment, a German biologist, August F. L. Weismann (1834-1914), cut off the tails of 1,592 mice at birth, over a period of twenty-two successive generations, and showed that all of them continued to give birth to mice with normal tails.

In the second place, many characteristics changed although they involved parts of the body that animals made no conscious use of. For instance, evolution would produce body coloring that would enable an animal to melt into the background and be, in this way, more secure from its enemies, yet it is inconceivable that a chameleon, for instance, would consciously *try* to alter its coloring and would therefore pass on a more efficient mechanism to its young.

In 1859, an English naturalist, Charles Robert Darwin (1809-1882), presented another suggestion, after gathering evidence on the subject over a period of fourteen years.

He suggested that in every generation a particular species contained members that differed from each other slightly in various characteristics, and were slower, faster, taller, shorter, stronger, weaker, redder, bluer, and so on. The many slight differences of varieties occurred at random, and individuals possessing one deviating characteristic or another might be more successful (on the average) in surviving than those possessing others.

The survivors are those who live and therefore transmit their special characteristics to their offspring, where again varieties exist that are, on the average, slower, faster, taller, shorter, stronger, weaker, redder, bluer, and so on. Again, some of the better-adapted varieties survive and breed so that the species over long periods of time become much slower, or faster, or taller, or shorter, or stronger, or weaker, or redder, or bluer. In different places or under different circumstances, different varieties prevail, so that a species begins to show two or more systematic variations that result at last in two or more different species. On some occasions, no varieties prevail, since none of them do as well as do certain different species altogether, and the first species becomes extinct.

Nature, in a sense, selects among the varieties that arise randomly, and this is "biological evolution by natural selection." It is this view of evolution that has prevailed. In the century and a quarter since Darwin, many refinements have been made in the theory, and there are continuing disputes over this detail or that. Nevertheless, although biologists may argue over the details of the mechanism of evolution, there are none of any standing at all who dispute the *fact* of evolution—just as a group of people may argue over exactly how a clock works, without ever disputing that it does, in fact, tell time.

Genetics

One of the points that Darwin left unclear was how the natural variations among members of a species could be involved in evolutionary development. Suppose that some members of a species were indeed a bit faster than others, and that speed was a valuable attribute in their case, one that contributed to better survival. Might not the fast members mate with slow ones (since organisms don't usually investigate the fitness of others before mating) and produce young of intermediate speeds? Would not, in fact, mating among organisms (which seems, to a great extent, to be random) generally lop off all extremes in properties and produce a vast gray intermediate, leaving nothing for natural selection to seize upon?

This was shown, in 1865, not to be so. An Austrian botanist, Johann Gregor Mendel (1822-1884), had carefully crossbred pea plants and studied what happened to the characteristics they displayed. For instance, he crossed pea plants with long stalks with pea plants with short stalks, and found that all the offspring had long stalks. He produced none with intermediate stalks. When he crossed these offspring among themselves, he found that of the new generation, some had long stalks and some short stalks, in a ratio of 3:1.

Mendel explained this by supposing that each plant had two factors of some sort that governed stalk length. Those with long stalks had two factors that helped produce length, and such plants might be referred to as *LL*. Those with short stalks had similar, but not identical, factors that produced shortness in stalk length, and we might use small letters to indicate this and label them *ll*.

On crossing the long stalks with the short stalks, each plant contributes *one* factor to each offspring, the factor being chosen at random. Whichever factor an *LL* plant contributes, that factor must be an *L*. The factor that an *ll* plant contributes must be an *l*. All offspring must have one of each and all are *Ll*, or *lL*. In this case, the *L* is "dominant," and the characteristic that it controls shows up. All the *Ll* and *lL* plants have long stalks, therefore, just as if they were *LL*.

The *l* factor has not disappeared, however, even though its effects seem to have. If *Ll* and *lL* plants are

217

crossed among themselves, each contributes an *L* to half the offspring and an *l* to the other half, purely by random choice. There are therefore four kinds of offspring: *LL*, *Ll*, *lL*, and *ll*. Of these, the first three have long stalks, the last has short stalks, and there's your 3:1 ratio.

Mendel showed that other sets of characteristics worked in the same way, and he carefully determined what are now known as the "Mendelian laws of inheritance." These showed that extremes are *not* lopped off in random matings but tend to persist and to show up, over and over, in later generations.

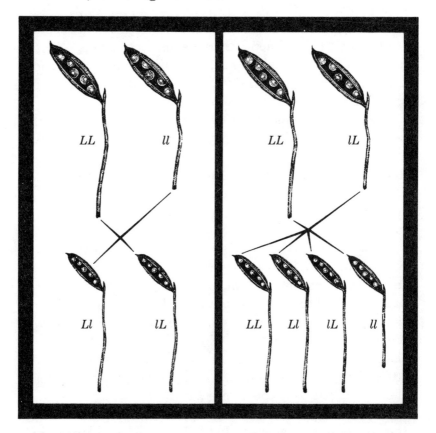

Mendel discovered the laws of genetics. They were so simple that anyone could understand them, but no one paid attention.

Unfortunately, Mendel was not well known as a botanist, and his work was ahead of its time. Although he published his experiments and his conclusions, he was completely ignored until 1900, when three other botanists figured out the same laws independently. All discovered that Mendel had been first by a generation, and each one, independently, gave Mendel full credit.

The greatest difficulty in Darwin's theory—the supposed tendency to lop off extremes—was thus solved.

Still, what was the biological and chemical nature of the factors that Mendel's laws required?

In 1882, the German anatomist Walther Flemming (1843-1905) reported on his studies of cells. He had developed techniques for exposing cells to some of the new synthetic dyes that chemists were developing. Certain dyes would combine with some features inside a cell and not with others. One dye in particular stained some of the material inside the nucleus. Flemming called this material *chromatin,* from a Greek word for "color."

It was known that the nucleus was essential to cell division, since a cell from which the nucleus had been removed would not divide. Therefore, Flemming stained a section of tissue in which the cells were actively dividing, and the chromatin was dyed in each. The staining killed the cells, but each cell was at some different stage of cell division so that there was a series of photographic "stills," so to speak, of the chromatin at the various stages. If these were put into what seemed the proper sequence, Flemming obtained a notion of the successive events in the process.

Apparently, as a cell divided, the chromatin collected into a group of stubby little rods that seemed to exist in pairs, so that there were two of each kind of rod. Flemming called each rod a *chromosome,* from Greek words meaning "colored body." The chromosomes lined up along the central axis of the cell and doubled, as though each

219

one produced another exactly like itself. Now there were two pairs, or four, of each chromosome.

The chromosomes then separated, two of each set of four moving to one end of the cell and the other two moving to the other end. The cell then pulled itself into two cells, each containing a full set of chromosomes arranged in pairs.

In 1887, the Belgian biologist Edouard Joseph van Beneden (1846-1910) studied chromosomes further. He showed that different species had characteristic numbers

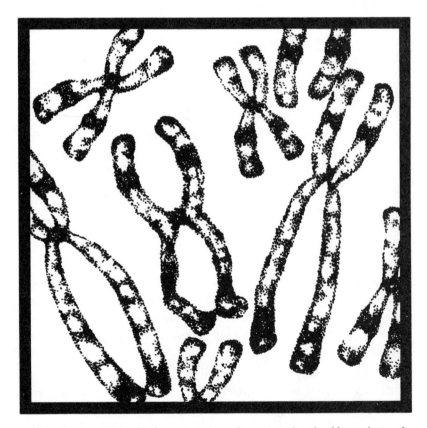

Chromosomes—the little structures that contain the blueprints of life.

of chromosomes in each cell. For instance, in every complete human cell (we now know) there are forty-six chromosomes arranged as twenty-three pairs. Beneden also showed that when an organism formed egg cells or sperm cells, those cells were each supplied with only half a set of chromosomes, one from each pair. (Thus, human sperm and egg cells each have twenty-three chromosomes.)

When a sperm cell fertilizes an egg cell, the fertilized egg cell once again has a full supply of chromosomes, one of each pair from the father and one from the mother. Thus, the fertilized human egg cell has twenty-three pairs of chromosomes.

In 1902, shortly after Mendel's work was rediscovered, an American biologist, Walter Stanborough Sutton (1877-1916), pointed out that the chromosomes behaved exactly like Mendel's factors and, indeed, had to be those factors. It was the chromosomes that controlled inheritance.

To be sure, there were too few chromosomes to explain inheritance if each chromosome was viewed as controlling one physical characteristic. Each chromosome, therefore, had to be viewed as being made up of a string of many molecules, each of which controlled one characteristic. In 1909, a Danish botanist, Wilhelm Ludwig Johannsen (1857-1927), suggested that these molecules be called *genes*, from a Greek word meaning "to give birth to." The study of these genes came to be called *genetics*.

10

NUCLEIC ACIDS AND MUTATIONS

Gene Structure

What are the genes? What kind of molecules are they?

The first hint of an answer had come in 1869, long before anyone but Mendel knew that genes existed. A Swiss biochemist, Johann Friedrich Miescher (1844-1895), had located a substance in cells that contained both nitrogen and phosphorus atoms. This substance was eventually called "nucleic acid" because it seemed to be located in cell nuclei.

Actually, it turned out that there were two varieties of nucleic acid. One was "ribonucleic acid" (abbreviated RNA) and the other "deoxyribonucleic acid" (DNA). It was the DNA that was, in the main, restricted to the nucleus and was, indeed, present in the chromosomes. RNA was found, generally, in the portion of the cell outside the nucleus.

Not much attention was paid to nucleic acid at first. It was thought to be a fairly simple compound, so small in fact that it could only have routine functions. On the other hand, the really important molecules in living tissue, scientists were certain, were the proteins, which exist in countless varieties and some of which are giant molecules made up of thousands of atoms.

Proteins are built up of amino acids, and there are twenty varieties of amino acids that can be put together in any fashion. We can imagine hundreds of amino acids jumbled together, including anywhere from one to thirty of each variety. Each different order in which these amino acids are put together would represent a different protein molecule with different properties. The mathematical number of possible different orders in which the amino acids can be joined is so enormous that there are far more different protein molecules that can conceivably exist than there are atoms in the universe—even if the entire universe were totally packed with atoms from end to end. If life is endlessly versatile and complex, it could only be (it seemed) because it must be based on the endlessly various protein molecule.

In comparison, the nucleic acid molecule was built up of units called "nucleotides." In any nucleic acid molecule there were only four different varieties of nucleotides, and it was for a long time thought that a nucleic acid molecule consisted only of four nucleotides all together, one of each variety.

Nucleic acids were first investigated in detail, from 1879 on, by a German biochemist, Martin L. A. Kossel (1853-1927). Kossel discovered much about the chemical structure of nucleotides, and he also found that sperm cells were particularly rich in nucleic acid (actually, in DNA, we now know) and that the protein present was considerably simpler than most proteins.

Since sperm cells have to carry all the characteristics inherited from the father, and are virtually nothing but tightly wrapped bundles of chromosomes, their composition had to be significant. Since they are loaded with DNA and rather light in protein, it would seem natural (in hindsight) to suppose that DNA and not protein was important in inheritance. However, the strong belief in the importance of proteins made it impossible for Kossel (or any other scientist at that time) to come to such a decision.

In 1937, the English botanist Frederick Charles Bawden (1908–   ) found that an example of the smallest form of life, a virus, contained nucleic acid as well as protein. Viruses are living organisms that consist of nothing but a nucleic acid molecule wrapped in a protein sheath (we now know).

All virus molecules seem to contain nucleic acid—some DNA, some RNA. (There are very small virus-like molecules called *prions* about which there is still some uncertainty in this respect.)

Considering that virus molecules are so simple, far smaller than cells, almost like isolated independent chromosomes, and are capable of multiplication once inside a cell, the presence of nucleic acid might seem to be significant. However, scientists, certain that proteins were of prime importance, simply assumed that it was the protein portion of viruses that were the essential working unit and that nucleic acids could only have some subsidiary function.

In 1944 came the turning point. In that year, the Canadian-American physician Oswald Theodore Avery (1877–1955) was studying two varieties of the bacterium that caused pneumonia. One contained a smooth coating around the cell and was called "S" for "smooth." The other did not, and therefore had a rough surface and was called "R" for "rough."

Apparently the R bacterium lacked a gene that would allow it to form a smooth coat. If S bacteria were killed and mashed up, however, something could be dissolved from the dead cell fragments, and this "extract," if added to the R bacteria, would cause the cells to begin forming a coat. The extract from the S bacteria must contain the gene that was missing in the R bacteria.

Avery and two co-workers purified the extract and removed as much as they could without removing whatever it was that gave the R bacteria the ability to form the coat. When they had finished the job, they found that the extract contained no protein at all, but it did contain nucleic acid. It was the nucleic acid, and not the protein, that was the gene.

By that time, it was beginning to be understood that nucleic acids, like proteins, were giant molecules made up of chains of hundreds, or even thousands, of nucleotides, distributed along the chain in any order at all. The only reason chemists had thought the nucleic acid molecules were simple was that they had been pulled out of the cells so roughly as to have been fragmented. Gentler procedures produced the intact molecule, and it then proved to be a giant.

At last scientists began to think about nucleic acids seriously, and about the DNA molecule in particular.

Two scientists, an Englishman, Francis H. C. Crick (1916–   ), and an American, James Dewey Watson (1928–   ), worked out the structure of DNA in 1953. They showed that the molecules consisted of two chains of nucleotides arranged in a double helix (that is, each one was shaped like a spiral staircase, the two curving upward parallel to each other).

The two nucleotide chains were held together by

chemical bonds between their atoms, and each one was shaped as the inverse of the other. In other words, where one bulged outward, the other curved inward, and vice versa, so that they fit together closely and tightly.

This solved the problem of how the DNA molecule forms a replica of itself ("replication") when chromosomes must form a new set in the process of cell division. The two nucleotide chains come apart (rather like an opening zipper) and each one serves as a mold onto which a new chain is formed. The new chain bulges out where

Within the chromosomes is the molecule of deoxyribonucleic acid (DNA), which is the blueprint of life.

the mold curves in, and vice versa. If you call the two chains A and B, then A serves as a mold on which to form a new B, while B serves as a mold on which to form a new A. The new chains form while the old chain is opening up, so that when the old chain is completely unzipped, the result is two chains, each as closely and as neatly zippered as the old one was.

Ever since 1953, scientists have been working out the details of the way in which the DNA molecule controls the cell. Although the DNA molecule is made up of only four different nucleotides, it is not the single nucleotides that are the key to control. The DNA molecule works through successive groups of three nucleotides ("trinucleotides"). Each trinucleotide can have any one of the four nucleotide varieties in the first position, any one of them in the second, and any one of them in the third. The number of different trinucleotides is, therefore, $4 \times 4 \times 4$, or 64.

Each trinucleotide corresponds to a particular amino acid. (There are more different trinucleotides than there are amino acids, so two or three trinucleotides may correspond to the same amino acid.) A particular section of the long DNA chain in a chromosome (that section making up a gene) can supervise the production of an amino acid chain corresponding to the chain of trinucleotides making up its own structure.

The protein formed in this fashion is an *enzyme*, and that has the ability to control the speed of a certain chemical reaction within the cell. All the genes in the chromosomes supervise the formation of all the enzymes in the cell. The nature of the enzymes, and the relative quantities of each, allow the cell to perform its characteristic functions, and, when all the cells are put together, we have a human being (or some other organism—depending on the nature of the genes).

Because the genes are passed from parents to offspring, the offspring are the same species as the parents and have the physical characteristics of the parents, so that not only do dogs give rise to dogs, but beagles give rise to beagles, and a particular pair of beagles will have puppies that will show the markings and other characteristics of the parents.

Gene Changes

But now arises a question. If DNA molecules replicate themselves exactly and are passed on from parents to offspring, why doesn't every organism have the same set of genes and, therefore, precisely the same physical characteristics?

Why and how have different species evolved? How is it that in a particular species, let's say beagles, there are variations in characteristics from litter to litter, and even within a single litter? Why do you look different from your brother or your sister?

The answer is that the replication of DNA isn't necessarily perfect. When the long nucleotide chain is molding another to itself out of individual nucleotides floating within the cell, every once in a while an inappropriate nucleotide is squeezed into position, and before it can be pushed loose, the chain is continued on either side and the wrong nucleotide is fixed in place. Chain A has thus produced a slightly misfitting chain B* (the asterisk shows a wrong nucleotide is in place). At the next replication, chain B* produces a new chain that fits itself, chain A*, and, thereafter, the misfitting DNA molecule remains part of particular members of that particular species.

Even a small change in a DNA molecule can alter the properties, sometimes quite noticeably. This means that offspring aren't invariably carbon copies of their parents. Sometimes, offspring have characteristics that belong to neither parent but can be traced back to earlier ancestors. And, sometimes, offspring have characteristics that you know did not belong to *any* of its ancestors.

People who keep herds of domestic animals know that sometimes animals are born with totally wrong colorings, or with abnormally short legs, or with two heads—or, in some other way, show features that are altogether surprising or new. Such offspring are called "sports," but scientists paid little attention to them.

In 1886, however, a Dutch botanist, Hugo Marie De Vries (1848-1935), who was later to be one of the three who rediscovered Mendel's theories, noticed a patch of flowers, all of the same species and all clearly born of seeds produced by a single bloom, which differed among themselves. He bred these plants and discovered that, every once in a while, offspring did not resemble parents in important particulars. He called these sudden changes *mutations*, from a Latin word meaning "change."

Once the method of DNA replication was understood, it was at once grasped that mutations were the result of imperfections in the replication procedure.

But why should there be imperfections? Well, nothing works perfectly all the time. When a new nucleotide chain is being put together, there is always the chance that the random collisions of molecules will result in wrong nucleotides bumping the wrong parts of the chain that is serving as mold. Usually, the wrong nucleotide doesn't stick and bounces away, but every once in a while, just by chance, the wrong one will happen to approach in such a way as to stick just long enough to be tied into the chain.

As an analogy, imagine a large party of people gathering for a meeting, with each person hanging up his coat in a cloakroom without an attendant. At the end of the meeting, everyone crowds into the cloakroom to get the coat that is his. Each person wants his own coat and knows more or less where he has placed it. On the whole, you would expect each person to emerge with his own coat. And yet, every once in a while, as you well know, someone will end up with someone else's coat, quite accidentally.

Mutations work on the same principle. Even though mutations take place only very rarely, there are so many replications among all the thousands of genes and all the billions of cell divisions that the total number of mutations is great. Perhaps every organism is born with a few mutations. These produce the variations in every generation of a particular species (although variations are also produced by differences in environment—in the quantity and kind of food available to young, in the presence or absence of disease or physical injury, and so on), and it is among these variations that natural selection can work to produce evolutionary changes.

Most variations, occurring as they do at random, are for the worse, that is, are disadvantageous. Thus, if you manage to pick up the wrong coat in the cloakroom, you are quite likely to find that it doesn't fit you, or that you don't care for the style. That "mutation" is for the worse, and you do your best to get your own coat back.

On the other hand, once in a very long while, you may find a coat that you actually like better than your own. Even if you have to give it back to the rightful owner, you may make up your mind to buy a coat like it, and thereafter you adopt that "mutation" and it becomes part of you.

Similarly, the mutation that takes place when a DNA molecule is imperfectly replicated may, on rare occasions, be beneficial in one way or another. It may help the organism to a successful or more adaptive life and to the production of many offspring, almost all of whom might inherit this mutation.

Even if there are 10,000 bad mutations to every good one, it is the good one that survives in more and more of the species while all the bad ones tend over time to die out. As a result, evolutionary change always seems to work in such a way as to make the species more successful.

We are not aware of all the changes that don't work and are gotten rid of. All we see are the very few useful changes. That is what makes it so hard to believe that evolutionary change is random and that there isn't a guiding intelligence behind it. If we could see *all* the changes, bad and good, it would be quite obvious that everything is working on a random basis and that it is the force of natural selection, choosing one out of many and rejecting the rest, that gives the illusion of purpose and direction.

It is the process of mutation, then—the very imperfection of DNA replication—that drives evolution forward and that has made it possible for human beings to come into existence. If it were not for mutation, if DNA replication were absolutely perfect, then once the first simple bit of life had been formed it would produce more bits exactly like itself, and that would be the end of it. All organisms that exist today would be replicas of that initial simple form of life.

And yet mutation by purely fortuitous circumstance does not occur often enough to account for the speed with which evolution has proceeded. Considering that it takes a million years or more for one species to evolve into another, you might not think evolution is a particularly speedy process, but it is, nevertheless, more rapid than mutation-by-chance would account for.

Since mutations take place more often than pure

chance would permit, there must be events on Earth that tend to increase the rate of mutation.

We can see this in our coats-in-the-cloakroom analogy. Suppose it becomes apparent that an unexpectedly large number of people are walking off with coats that are not theirs. There must be factors that are increasing the mistake rate. One of the cloakroom lights may have gone out, and in dim light it is less easy to choose correctly among similar coats, so that mistakes would happen more often. Or it might be that everyone has had a number of drinks. With woozy vision and defective judgment, the rate of mistakes would increase. A third possibility might arise if there were a crisis. Once the people were all in the cloakroom, a shout of "The bus is leaving!" would make everyone grab quickly and the number of mistakes would again increase.

Mutagenic Factors

Something that would make the rate of mutations rise could be called a "mutagenic factor," or, more briefly, a *mutagen*, from Greek words meaning "to give rise to change." What are the mutagenic factors that could increase the rate of mutation and thus produce evolutionary change at the observed speed?

One such factor is a rise in temperature. The higher the temperature, the more rapidly atoms and molecules move and jiggle, and the harder it is to select the right one from a crowd. The mutation rate would go up as the temperature went up.

Life, however, developed in the ocean and stayed in the ocean until about 400 million years ago. For nearly nine-tenths of its existence on Earth, in other words, life was to be found only in the ocean.

As it happens, the ocean environment is much more stable than the environment on dry land. The temperature in the ocean doesn't change much from season to season and year to year (certainly not as much as temperature on land does). Through almost all the history of life, then, the effect of temperature on mutations has been small and cannot be considered as making evolution possible at its observed rate.

There are also chemicals that act as mutagens because they tend to combine with DNA and thus produce abnormalities by their presence when replication takes place. Or else they react with DNA in such a way that even though they do not combine with it, they do change the arrangement of some of the atoms making up the molecule. A DNA molecule with an abnormal set of atom arrangements will serve as an abnormal mold during replication and will produce a mutation.

However, organisms that are easily affected by the chemicals they are likely to encounter are so overwhelmed by mutations (almost all of them for the worse) that they quickly die out. The force of natural selection chooses those that, for one reason or another, are resistant to chemical mutagens, so that in the end we needn't expect much from them as a way of speeding evolution.

Nowadays, of course, mutagens have become a serious problem. Chemists have produced many thousands of new compounds that do not exist in nature and that have been put into the environment in considerable quantity. Some of them are mutagens, and organisms have not had a chance to encounter them before so that they have not yet, by natural selection, developed resistance to them. As a result, many organisms (including human beings) can be adversely affected by them.

Some mutations, for instance, change normal cells into cancerous cells by the production of an abnormal *oncogene*, where "onco" is from a Greek word for an abnormal growth, such as cancer produces. Mutagens that bring about such a change are called *carcinogens*, from a Greek word for "crab," because a cancer sometimes spreads out in all directions like a crab's legs.

Still, through all the billions of years of life before this last century of chemical development, mutagens were not much of a problem and cannot be relied on as an explanation of the rate of evolutionary change.

A mutagenic factor that is much more efficient than heat or chemicals was first noted by the American biologist Hermann Joseph Muller (1890-1967). He was working with fruit flies and waiting for chance mutations so that he could study the ways in which those mutations were inherited. Waiting for random mutations, however, was too tedious and time consuming, and he looked for ways to speed up the mutation rate.

In 1919, he raised the temperature in which his fruit fly colonies lived, and the mutation rate did go up, but not enough.

It occurred to him to try x-rays. They were more energetic than gentle heat and would penetrate the fruit flies from end to end. If, in passing through the fruit flies' bodies, an x-ray should happen to strike a chromosome, it would be energetic enough to knock atoms here and there. This would inevitably induce a chemical change—in other words, a mutation. Muller did not know what the chemical nature of the genes was (no one was to know until thirty years later), but whatever that nature might be, he was sure the x-rays would effect changes.

He was right. By 1926, he could show beyond doubt that x-rays greatly increased the mutation rate.

Others began to investigate this new effect, and it turned out that any kind of energetic radiation would raise the mutation rate. Ultraviolet rays would do so, and so would the radiation from radioactive substances.

Yet, how could energetic radiation be responsible for the mutation rate that forced evolution to proceed as rapidly as it did?

X-rays have been produced by human technology over the last century, but prior to that there has been little in the way of x-rays on Earth. To be sure, the Sun's corona radiates x-rays constantly, as do other objects in the sky, but these are largely absorbed by our atmosphere and do not reach Earth's surface.

Radioactive substances certainly exist on Earth and existed in perhaps twice the quantities during the infancy of life on this planet. However, they exist mostly in the soil, and sea life would not be greatly affected by them. Even on dry land, they are not distributed evenly, and there are few places on Earth where natural sources of radioactivity are sufficient to be an important source of mutations.

Ultraviolet light from the Sun is, in a way, less of a danger than x-rays or radioactive radiations, since ultraviolet light is less energetic than either of the other two. On the other hand, ultraviolet light is always present in sunlight, especially in those early ages when the ozone layer in the upper atmosphere had not yet been formed.

Sunlight, with its ultraviolet, would appear inescapable. Ultraviolet is energetic enough, in the quantities and wavelength range that were present before the days of the ozone layer, to produce not only mutations but also the kinds of chemical changes that might kill living organisms outright. It may have been for that reason that it took so long for life to colonize the dry land. Until enough of an ozone layer had formed to block the more energetic portions of the solar radiation, emergence onto dry land in the full blaze of sunlight could have been fatal.

Ultraviolet light, however, is more efficiently absorbed by water than by air. Ocean life would have evolved the kind of behavior that would allow it to sink a number of feet below the water surface when sunlight is shining directly down upon that surface. Ocean life could rise to the surface when the Sun is near the horizon (or below it) and when the day is cloudy. After plant cells evolved and sunlight became essential to their functioning, those cells might still sink to a level that would allow them to receive enough radiation for photosynthesis to continue, but not enough to represent a mortal danger. And, of course, once plant cells evolved, an oxygenated atmosphere and an ozone layer in the upper reaches soon came into being, and the ultraviolet danger largely disappeared.

But since all mutagenic factors mentioned in this section seem relatively ineffective, how, then, do we account for the observed rate of evolution? To find the answer, let us take a new approach.

Cosmic Rays

After radioactive radiations were discovered in the last decade of the nineteenth century, scientists developed devices to detect such radiations. Rather to their surprise, they discovered that even when radioactive substances were nowhere about (so far as was known) some radiation somewhere was being detected by their devices. What's more, even when the devices were hedged about with lead shields that would be opaque to radioactive radiations (and to all other forms of radiation then known), the devices *still* detected radiation.

Apparently, radiation existed that was not only of an unknown origin, but that was more penetrating (and hence more energetic) than any of the known kinds. It was more energetic even than the gamma rays that were emitted by certain radioactive substances, and the gamma rays were, in turn, more energetic than x-rays.

It was assumed that this new form of radiation had its source in some substances in the soil, some super-radioactive substances, but this was only an assumption. It occurred to an Austrian physicist, Victor Franz Hess (1883–1964), that this might be confirmed rather easily by taking radiation-detecting instruments up into the air in a balloon. The farther these were carried above the Earth, the weaker the radiation should become, assuming the source were really in the soil.

Beginning in 1911, Hess made ten balloon ascents with his instruments, five by day and five by night. One of the daytime ascents was during a total eclipse of the Sun. He found, to his astonishment, that he hadn't demonstrated the soil to be the source at all. In fact, quite the reverse. The higher he went by balloon, the more intense the penetrating radiation became. The source was apparently in the sky and not in the soil. What's more, it wasn't the Sun that was responsible, for the radiation intensity remained the same whether it was daylight or not.

As nearly as Hess and others could make out, the radiation came equally from all parts of the sky. The American physicist Robert Andrews Millikan (1868–1953) called the radiation "cosmic rays" because it came from the cosmos generally, and the name caught on. Millikan believed that cosmic rays were another kind of electromagnetic radiation, like ordinary light.

Electromagnetic radiation behaves as if it consists of waves. The tinier the waves (that is, the shorter the wavelength) the more energetic the radiation. Visible light has very short waves to begin with, and, of the different colors of light, red light has the longest waves and is least energetic. The wavelength grows shorter and the energy content higher as one passes through the spectrum of red, orange, yellow, green, blue, and, finally, violet light.

Ultraviolet light has wavelengths shorter than that of violet light so that it is more energetic than any visible form of light. X-rays have shorter wavelengths still, and gamma rays even shorter. It was Millikan's view that cosmic rays were ultra-short gamma rays and were, therefore, more energetic and more penetrating than gamma rays.

This view was disputed by another American physicist, Arthur Holly Compton (1892–1962), who felt that cosmic rays must consist of very speedy electrically charged subatomic particles. Their energy rested in their momentum, which depended on both their mass and speed.

There was a way of possibly settling this dispute.

If cosmic rays were electromagnetic radiation, they would have no electric charge and would be unaffected by Earth's magnetic field. They would therefore strike different places on Earth in identical fashion, for they were apparently coming equally from all parts of the sky.

On the other hand, if the cosmic rays were electrically charged particles, they would be affected by Earth's magnetic field. They would be deflected toward Earth's magnetic poles. These cosmic ray particles (if that was what they were) were so energetic that they would be only slightly affected and only slightly deflected. Compton, however, calculated that the amount of deflection should be detectable and that, in general, the further one traveled from the equator, either north or south, the more intense the cosmic ray bombardment should be.

Beginning in 1930, Compton became a world traveler in order to check his supposition and was able to demonstrate that he was correct. A "latitude effect" did exist, and cosmic ray intensity was greater, the higher the latitude. Millikan held out stubbornly, but gradually the world of physics swung behind Compton. The particle nature of cosmic rays is now well established.

Cosmic rays consist, as is now known, very largely of positively charged subatomic particles, mostly hydrogen nuclei and helium nuclei in a 10:1 ratio. There is also a scattering of more massive nuclei, all the way up to some iron nuclei. The distribution of nuclei in cosmic rays is similar to the distribution of elements in the universe.

It is no surprise that cosmic rays are so energetic and penetrating, for the particles move much more rapidly than similar particles that develop on or near Earth, including those originating from radioactive substances. The most energetic cosmic ray particles travel at a speed only slightly less than the speed of light—which is the absolute maximum for anything possessing mass.

The existence of cosmic ray particles has an important relationship to biological evolution. These particles, being energetic, can and do cause mutations.

Cosmic ray particles do not strike the Earth in quantities at all comparable to the ultraviolet of sunlight, or the x-rays from an x-ray generator, or the radiation from nearby radioactive substances. However, although one can avoid going near x-ray sources or radioactive substances, and even avoid the ubiquitous ultraviolet by something as simple as remaining in the shade, there is no reasonable way of avoiding cosmic ray particles.

One might, to be sure, go into a mine well below the surface of the Earth, or live in an air bubble at the bottom of a deep lake, or surround oneself with a thickness of several feet of lead—but the vast majority of living things do not follow, and never have followed, any of these strategies.

For billions of years, living organisms must have encountered little in the way of energetic electromagnetic radiation, radioactive radiation, or mutagenic chemicals, but they were constantly bombarded with cosmic ray particles day and night, wherever they were. The atmosphere and water, which blocked much of the ordinary radiation from the Sun and from the sky generally, did not block the cosmic ray particles.

To be sure, the cosmic ray particles as they existed in space ("primary radiation") were not unchanged. They struck atoms and molecules of Earth's atmosphere and were slowed and absorbed. In the process, however, they knocked energetic particles ("secondary radiation") out of the atoms and molecules, and these, in one form or another, still intensely mutagenic, reached the Earth's surface and penetrated deeply into soil and water.

It can be concluded that the steady bombardment of cosmic ray particles that bathed life through all its existence must have been mild enough to allow organisms to live comfortably, but it was intense enough to raise the mutation rate well above what it would have been if it were only a matter of accidental imperfections in the replication process, or if it were only the added push of mutagenic factors that were less common or more avoidable than cosmic ray particles.

It would seem to be cosmic ray particles, then, more than anything else, that powered mutations which, in turn, gave natural selection a handle and made evolution proceed at the speed it did. It is to cosmic ray particles that we owe our own existence, for at the speed of evolutionary change without them, it might well be that life on Earth might still consist of nothing more complex than worm-like creatures living in the sea.

But where do cosmic rays come from?

Since they come from all over the sky, they can't be

pinned down to any one object, or to several specific objects here or there. Nor can individual bursts of cosmic ray particles be supposed to come from some object in the sky that was near the point from which they seemed to radiate.

Electromagnetic radiation travels in virtually a straight line (except for very slight curvatures when it passes quite close to a massive object). That means that if you see a ray of light, the source is from the direction in which you are looking when you see it. If you see a star by the light it emits, you find yourself looking at the star itself when you are looking at the light. People are so used to this straight-line propagation of light that if you say, "A star is where you see it to be," it sounds like a totally unnecessary statement. Where else would it be?

Any other form of electromagnetic radiation originates in the point of the sky from which it appears to come. We take that for granted, too.

However, electrically charged particles do *not* travel in a straight line. They are affected by magnetic fields, and the galaxy is full of magnetic fields. Every star has one, so do many planets, and the galaxy as a whole has one. A cosmic ray particle, therefore, as it streaks through interstellar space, follows a very complex curved path as it responds to all the magnetic fields it passes through.

As it zooms down to Earth's surface, the direction it follows in this final portion of its journey is no indication of its path when it was a dozen light-years away. It is as though you looked at a bird or a bat coming at you in a line that, if you traced it backward, would end in a distant tree. That is no sign that the bird or bat came from that tree, since it may have veered unpredictably a dozen times in the course of its journey.

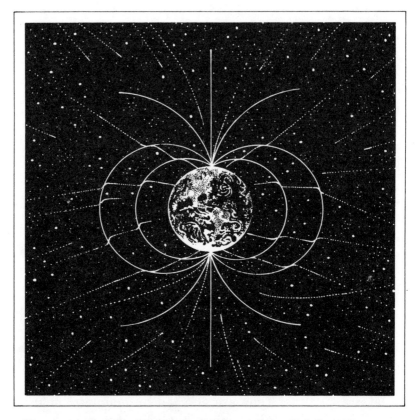

The magnetic field of Earth is even less substantial than the ozono-sphere, and is also a powerful protecting force.

With every cosmic ray particle following its own complex path, it's no wonder that they seem to be coming from every part of the sky, and neither is it any wonder that you can't trace any of them back to a source.

However, we do know that cosmic ray particles are enormously energetic and that, wherever they come from, their source must be something violent. You don't get energetic particles from some quiet process.

The most violent object in the solar system is, of course, the Sun, and the most violent event on the solar surface is a solar flare. Is a solar flare violent enough to produce cosmic ray particles?

The question was not really asked, but the answer came anyway and forced itself on scientists.

Toward the end of February 1942, a large solar flare appeared in the center of the Sun's face, which meant it was shooting material directly toward the Earth. Very soon afterward, a burst of comparatively weak cosmic ray particles was detected. The direction of its approach was from the Sun, and that could be accepted as showing the Sun to be the source, for in the paltry distance that separated Sun and Earth, there was no time or occasion for the speeding cosmic ray particles to change their direction of travel measurably.

Since then, numerous bursts of "soft" cosmic ray particles have reached us after large flares appeared in appropriate positions on the face of the Sun.

There is no mystery about this, now that the fact is known. The solar wind is a stream of outward-hurling nuclei, mostly hydrogen and helium. These nuclei are energetic, traveling at hundreds of kilometers per second. Solar flares are particularly energetic events on the Sun's surface, and they produce a gust of solar wind in which the particles travel at much greater speeds. If the flares are energetic enough, and the wind speedy enough, the particles are cosmic ray particles.

Cosmic ray particles are of the same order of objects as solar wind particles; the only difference lies in the greater speed and energy of the former—just as the only difference between x-rays and light waves is the shorter waves and greater energy of the former.

At best, though, the Sun is only capable of emitting occasional bursts of cosmic ray particles at the lowest range of energies. To produce more energetic cosmic rays in quantities huge enough to fill the galaxy, events are needed that are far more violent than the Sun's quiet middle-aged nature can produce.

Clearly, then, the most violent stellar events are supernova explosions, and it makes sense to suppose that each such explosion sends forth a vast burst of an incredibly energetic stellar wind in all directions. These are cosmic ray particles.

These particles travel through the near-emptiness of interstellar space without slowing down. Indeed, as they whip about curves in response to the magnetic fields they pass through, they may accelerate ever nearer to light speed. The more energetic they become, the less they veer from their straight path in response to magnetic fields, and, eventually, nothing can veer them enough to keep them from speeding out of the galaxy altogether and into the still-emptier intergalactic spaces.

This is not the fate of all cosmic ray particles. A number of them, in their long journey, are bound to strike other pieces of matter, whether the occasional atom or dust grain in interstellar space, or a star, or something in between, such as our own Earth.

There are enough cosmic ray particles in space released by all the supernovas that have ever exploded in the course of the galaxy's history to insure that a sizable number of them strike the Earth every second, coming from all directions. To be sure, a certain percentage of the cosmic ray particles produced by our galactic supernovas escape our galaxy, but these are balanced by the number that must reach us from other galaxies.

And so it is that not only did supernovas supply the raw material out of which the Earth and living matter were formed; not only did they supply the heat that kept the cloud out of which our solar system formed from having condensed prematurely; not only did they supply the pressure shock that made the condensation eventually possible; but they also supplied the driving force behind the evolutionary changes that converted life on Earth into more and more complex forms and, eventually, into human beings.

Supernovas, then, are titanic crucibles in space—immense anvils, whose workings hammer out matter and whose products create the surroundings that permitted life, at least once, to form and evolve.

11

THE FUTURE

Earth's Magnetic Field

So far, the effects of supernovas on mankind, as I have described them, would seem to be entirely benign. Is it possible, however, that supernovas might work to our disadvantage in some ways and at some times? Might they even threaten the existence of humanity on occasion? Or all of life?

Clearly, a supernova, if it exploded nearby, galactically speaking, could deliver energy in killing intensities. If our own Sun were to go supernova, for instance, not only would all life on Earth come to an end within minutes, the globe of the Earth itself would vaporize. If the Sun were merely to go nova, for that matter, Earth would probably be sterilized.

As has been emphasized earlier, however, this cannot happen. Our Sun is not very massive and is not part of a close binary system, so there is no possible way, now or ever, in which it can go nova or supernova. It will eventually become a red giant and then undergo collapse to a white dwarf, but, until that happens (five or six billion years from now) nothing will happen to the Sun (barring something as unlikely as a collision or near-collision with another star) that will be threatening to life generally.

Might we be damaged if stars other than the Sun were to explode? Even the nearest stars thought to be capable of supernova formation are over 100 parsecs away. If any of them exploded tomorrow, it is just possible there could be some deleterious effects, but nothing, perhaps, over that vast distance that would really endanger humanity as a whole.

The nearer supernovas of the past have not, after all, affected us. The supernova that produced the Crab nebula did not, and even the Vela supernova that was close enough to shine as brightly as the full Moon for a few days in prehistoric times did not interfere with life on Earth, so far as we know.

The one direct effect upon us of a distant supernova that is strong enough to be significant is that of the cosmic rays it produces, so let us consider cosmic rays once again.

The total energy delivered to Earth by cosmic rays is surprisingly large. The energy is believed to be roughly equal to the total energy delivered to us by the light of all the stars in the sky, exclusive of our Sun. The number of individual cosmic ray particles is much smaller, to be sure, than the number of light photons that reach us from the stars, but the individual cosmic ray particle is far more energetic than the individual photon and that makes up for it.

By and large, the incidence of cosmic ray particles on Earth is quite steady (barring the occasional and temporary wash of comparatively feeble particles of that kind resulting from a particularly energetic solar flare), but suppose that, for some reason, that incidence were to increase markedly for a period of time. Could this do harm?

The answer is yes!

Cosmic ray particles produce mutations, which are necessary if evolution is to progress at a reasonable speed, but most mutations are, nevertheless, harmful. Fortunately, thanks to natural selection, the few mutations for the better survive and spread, under ordinary conditions, while the mutations for the worse die out. Even so, the mutations for the worse do produce a "genetic load" on the species, a certain percentage of the population that is relatively unfit for survival.

What if conditions are *not* ordinary, however? What if the cosmic ray intensity increases to well above the normal level and remains there for a while? The mutation rate would increase and so would the genetic load. It is possible for the genetic load to become so heavy that the population of the species would decline precipitously, the few beneficial mutations would not be able to retrieve the situation, and the species would become extinct. A number of species might become extinct at more or less the same time.

But can the level of cosmic ray intensity increase for any reason *other* than the appearance of supernovas near us in space?

Oddly enough that level *can* increase, and we may, in fact, be facing an inevitable increase over the next couple of thousand years, even if no supernovas appear to supply additional cosmic rays. To explain, let us backtrack a little.

248

Not all cosmic ray particles that approach the Earth actually hit it. The Earth has a magnetic field, something known since the days of the English physicist William Gilbert (1544-1603), who in 1600 published a book describing his experiments with a sphere of magnetic material. A compass needle in the neighborhood of that sphere acted exactly as it did in the neighborhood of the Earth, which implied that the Earth, too, was (in a way) a sphere of magnetic material.

If you imagine continuous lines drawn through Earth's magnetic field connecting points of equal magnetic intensity, you have a family of "magnetic lines of force." All of these start and end at two points on Earth's surface, one at the rim of Antarctica (the "South magnetic pole") and one at the northern rim of North America (the "North magnetic pole"). In between, they belly upward in smooth curves that follow a more or less north-south direction, reaching their highest point halfway between the magnetic poles.

Any electrically charged particle that hurtles from outer space to Earth's surface must cross these magnetic lines of force, and that takes energy. The particle loses energy and is slowed as it does so. Furthermore, an electrically charged particle that is not aimed directly at Earth's magnetic equator is deflected in such a way that it tends to bend in the direction of the magnetic lines of force, bending northward when north of the magnetic equator and southward when south of it.

The less energetic a particle is, the more it is deflected, and if such a particle is sufficiently low-energy, it is trapped by the magnetic lines of force and is forced to slide along them and eventually move into the atmosphere at or near the sites of the magnetic poles.

Cosmic ray particles are so energetic that they are deflected only slightly by Earth's magnetic field. However, some of these particles that would otherwise just strike the rim of the circle of the Earth would be sufficiently deflected to miss it altogether. Even those that are approaching more directly are deflected to some extent. For this reason, many of the cosmic ray particles that would ordinarily strike in the tropic and temperate zones, where Earth is rich in land life, end up striking in the polar zones, where Earth is poor in land life.

In this way, the effect of cosmic ray particles on life is somewhat reduced by Earth's magnetic field. It is reduced to an extent that lowers their potential for doing damage, but not to an extent that prevents them from performing their useful purpose with respect to evolution.

The weaker the Earth's magnetic field, the less it will serve to deflect the particles and, therefore, the greater the intensity of cosmic rays on Earth's surface, particularly on the lower latitudes.

As it happens, the Earth's magnetic field does not remain at constant strength. Since scientists began making measurements back in 1670, the magnetic-field intensity has declined some 15 percent. If the field were to continue to decline in intensity, it would reach zero at about the year 4000.

But will it continue to decline in intensity? That doesn't, at first thought, seem likely. It would seem much more probable that the intensity fluctuates, becoming weaker and weaker until it reaches some still fairly-high minimum, then becoming stronger and stronger until it reaches some not-extremely-high maximum, then repeating the process over and over.

It might seem that the only way we can tell what will really happen would be to wait for a few thousand years and continue to measure magnetic-field intensity; but, as it turned out, it is not necessary to do this.

There are certain minerals in the Earth's crust that have weak magnetic properties. When lava from volcanoes cools and solidifies, such minerals form crystals that line up north and south in the direction of Earth's magnetic lines of force. What's more, each crystal has a north pole that points north and a south pole at the opposite end that points south. (One can tell which pole is "north" and which "south" by testing the crystal with an ordinary magnet.)

In 1906, the French physicist Bernard Brunhes (1869-1930) was studying volcanic rocks and noted that in some cases the crystals were magnetized in the direction *opposite* to Earth's present magnetic field. The north poles faced south, and the south poles faced north. The finding was ignored at first, because it seemed to make no sense, but, with time, additional evidence accumulated and now the matter can neither be denied nor ignored.

Why are some rocks oriented the "wrong" way? Clearly, because the Earth's magnetic field points sometimes one way and sometimes another. Rocks that cool and crystallize when the Earth's magnetic field points in one way also point in that way. When the magnetic field reverses itself, however, it lacks the strength to force the crystals to reverse their direction. The crystals now point the wrong way.

In the 1960s, the magnetic properties of the sea floor were studied. The bottom of the Atlantic Ocean, for instance, has spread outward to its present width by the welling up of molten material from the Earth's interior through a long, curving rift that runs down the central line of the ocean. The rocks near the rift are the newest and the most recently solidified. As one progresses away from the rift in either direction, the rocks are older and older. If the magnetic properties are studied, the rocks nearest the rift point the "right" way, in line with the present direction of the magnetic field. Further away from the rift, they point the wrong way, then the right way, then the wrong way. There are stripes of right and wrong on either side of the rift, one side a mirror image of the other.

By measuring the ages of these rocks, it turns out that the magnetic field reverses itself at irregular intervals. Sometimes there are only 50,000 years between reversals, sometimes as much as twenty million years. Apparently what happens is that the magnetic field periodically declines in intensity to zero and then continues to decline "below zero"; that is, reversing its direction and becoming more and more intense in that direction. It then declines to zero again, reverses again, and so on.

What causes the magnetic field to rise and fall in intensity in so irregular a fashion and to reverse its direction with each passage through zero? Scientists do not yet know, but they feel certain that it happens just the same.

Right now, we seem to be headed toward such a reversal, which will take place, as I said earlier, about the year 4000. For a few centuries, before and after, the magnetic field will be so weak that it will not serve to deflect cosmic ray particles to any significant extent.

As the magnetic field rises and falls, the cosmic ray incidence falls and rises. The cosmic ray incidence falls to a minimum when the magnetic field is most intense and rises to a maximum when the magnetic field reaches zero intensity.

When the magnetic field is at zero intensity, and the cosmic ray incidence is at maximum, the mutation rate and the genetic load are also at maximum. It is then that the chance for some species to become extinct will be at its peak.

The Great Dyings

Species have become extinct throughout the history of life on Earth, to be sure, but it hasn't happened in any even way. As paleontologists studied the fossil record, they became aware of the fact that there have been periods when extinctions have been unusually high, even periods when it seemed that the majority of living species seemed to have grown extinct over a comparatively short period of time.

These periods are sometimes called, rather dramatically, "The Great Dyings." The best-known such period took place about sixty-five million years ago, at which time the huge reptiles that dominated the Earth, including the many creatures we call "dinosaurs" together with numerous other species of organisms, were all hurled into extinction.

Could these Great Dyings correspond to periods of zero-intensity magnetic fields? Are we heading for such a Great Dying in A.D. 4000, and will humankind be wiped out not long after?

Apparently, this is *not* something we need fear. We can't trace magnetic field reversals back in time for very many millions of years, but we know that a number have occurred in the last few tens of millions of years and that these have *not* necessarily been accompanied by unusual numbers of extinctions. We need not expect, therefore, a genetic-load disaster in about 2,000 years.

Nor is this surprising. The Earth's magnetic field is not very strong, at best, and cosmic ray particles are exceedingly energetic so that deflection is not very great. Therefore, as the magnetic field intensity falls, and the cosmic ray incidence rises, it does not rise very much simply because it doesn't fall very much in the first place when the magnetic field intensifies.

But what if the cosmic ray intensity were to rise without reference to Earth's magnetic field? What if a supernova were to explode nearby? There would then be a temporary flood of cosmic ray particles striking Earth, and this might account for the many extinctions.

Imagine a large supernova exploding no more than ten parsecs from Earth. It might briefly shine with a light perhaps $\frac{1}{600}$ that of our Sun, and would thus be far brighter than anything else in the sky, including the Moon. If it were on the opposite side of the Earth from the Sun, it would turn night into a kind of twilit day. No matter where it was in the sky, it would warm the Earth appreciably for a while and make things uncomfortable for us all.

More importantly, cosmic ray intensity would be increased to hundreds or even thousands of times what it is now, and this rise might continue to be very substantial for many years. There could be all kinds of unpleasant consequences. The ozone layer might be weakened so that more ultraviolet light might reach the surface, and this could be as deadly to life as the cosmic ray particles themselves. Some of the nitrogen and oxygen in the atmosphere might combine, and the nitric oxide that would form in the upper regions would block some of the visible light. Temperatures would then drop after the initial rise and so would the precipitation rate. And there would, of course, be a great rise in the mutation rate and the genetic load.

If all this happened just when Earth's magnetic field was at or near zero, the effects would be slightly intensified and all the worse for that. Could it be that the Great Dyings were the result of a combination of a nearby supernova and the temporary disappearance of the field?

There are no stars within ten parsecs of ourselves that can possibly go supernova so, at first thought, the

254

suggestion seems a ridiculous one. The Sun, however, is moving, and so are all the other stars in our galaxy. The motions carry the stars about the center of the galaxy, but they don't move in chorus-line unison. Those further from the center move more slowly than those closer to the center. Some (like the Sun) have nearly circular orbits, while others have highly elliptical ones. Some move in the general plane of the Milky Way, others move in planes at a considerable angle to that.

The result is that stars may approach other stars, then recede from them, approach still others and recede; and this may happen over and over again during each orbit about the galactic center. The chances of actual collision are exceedingly small, but to pass within ten parsecs of another star is not uncommon. We are within 1.3 parsecs of Alpha Centauri, and within 2.7 parsecs of Sirius. We weren't always at those distances, and we won't always be in the future.

Is it possible, then, that at various times in the Sun's long past it did move fairly close to a star just when it happened to go supernova, and that it may happen at various times in the future? Could such events account for the Great Dyings and, in particular, for the disappearance of the dinosaurs?

In the late 1970s, this suggestion gained considerable popularity among scientists.

In 1980, however, the American physicist Walter Alvarez discovered that in a rock layer sixty-five million years old there was a surprisingly high quantity of the rare metal iridium. He suggested that a large asteroid might have struck the Earth some sixty-five million years ago and kicked so much dust into the stratosphere as to block all sunlight from Earth for a considerable period of time. This would result in the Great Dying that killed off the dinosaurs. The dust eventually settled out over the face of the Earth, bringing down with it a fine powder of iridium in which the original asteroid had been comparatively rich.

Since then, considerable supporting data have been uncovered to substantiate this suggestion. What's more, evidence was collected in 1983 to show that Great Dyings occurred with unexpected regularity—every 26-28 million years. Astronomers had to consider what factors might be responsible for such a long-scale periodicity.

They speculate, for instance, that the Sun might have a distant companion, not quite large enough to shine as a star, which in part of its twenty-seven-million-year orbit approaches closely enough to the Sun to pass through a cloud of hundreds of billions of comets that may be circling in orbits far beyond the planet Pluto. Hundreds of thousands of those comets would, by the gravitational pull of the companion, be deflected into orbits that would carry them into the inner solar system. A few would be sure to hit the Earth and cause the havoc that would bring on mass extinctions.

The last Great Dying took place about eleven million years ago, and if the suggestion of cometary impacts is correct, the next one will take place about sixteen million years from now. There is no need for immediate concern.

Supernovas would now seem (pending further shifts in evidence and interpretation) to be absolved of responsibility for the Great Dyings. Nevertheless, it remains possible that an occasional relatively-nearby supernova may produce enough cosmic radiation to produce extinctions that would not otherwise take place.

Space

In the future, there will be specialized conditions under which cosmic rays are bound to be of far greater concern than they are to us now.

Consider space travel, for instance. Already, a number of human beings have been in near space, outside all but the thinnest wisps of the upper atmosphere. In some cases, they have moved outward as far as the Moon.

An astronaut orbiting the Earth is outside the protective layers of the atmosphere, but he still has the planet's magnetic field between himself and the streams of cosmic ray particles arising from the Sun and from other sources in outer space.

So far, astronauts have shown no apparent harm from exposure to the conditions of space. Even those Soviet cosmonauts who have remained in orbit for up to eight months at a time seem to have survived quite well. (One of them, in two separate tours of duty, has remained beyond the atmosphere for a year.)

An astronaut, traveling to the Moon and back, is beyond Earth's magnetic field as well as its atmosphere, and the Moon itself has no perceptible amount of either. Such astronauts are therefore exposed to the full intensity of cosmic ray bombardment for as long as six days, and they have shown no ill effects as a result.

Nevertheless, the time will come when there will be still more in the way of exposure. Spaceships, with crews aboard, will perhaps make their way to Mars and beyond in times to come. Exposure to cosmic ray bombardment will then continue not for a matter of days but for months or even years.

In addition, there is the possibility of space settlements that will be occupied by thousands of human beings for indefinite periods. We will be talking, then, not of years but of lifetimes and generations. We will be facing a time when children are conceived in space, born in space, brought up in space. Will cosmic ray bombardment increase the rate of mutations? Will there be an increase in birth defects? Will the increasing genetic load make life in space difficult or impossible?

If space settlements are large enough, they can be shielded from cosmic rays, at least in part, even without a miles-thick atmosphere and a planet-wide magnetic field to do the job.

The settlements will (it is probable) be built out of metal and glass obtained by mining the Moon. Moon rocks, pulverized, will also form the soil that will be layered onto the inner surface of the settlement, and that will be held there by centrifugal effect as the settlement rotates. The soil will be the basis of farming activities in

Space settlements may in the future represent a major home of humanity.

the settlement, but it can also be made thick enough to absorb a large fraction of the cosmic ray particles.

Truly long space flights will be made with large spaceships, built and launched in space, that can be designed as small worlds in themselves. They, too, can be coated with soil inside the hull, both for growing their own food and for serving as a cosmic ray absorber.

There will come times, however, when the cosmic ray danger increases temporarily. Every once in a while, a giant solar flare may yield a spurt of cosmic ray particles that will wash over space settlements and spaceships. Such a spurt will perhaps not last for long and will yield rather feeble particles, by cosmic ray standards. No doubt the protective layers of soil will take care of them.

The unexpected explosion of a supernova will also add to the cosmic ray flux. It will do so much more rarely, but will provide much more energetic particles over a much longer period. However, such supernovas will usually be far enough away to be of little danger.

Of course, one can always imagine a combination of events that might bring about tragedy. Once we have space settlements and a space-centered society, there are bound to be people who, at any given time, are making short trips from settlement to settlement in small and unshielded shuttlecraft, or who are working in space with no more protection than a spacesuit. If there should be a sudden unusual wash of cosmic ray bombardment, whether from the Sun or from a supernova, significant damage might be done and, occasionally, lives might be considerably shortened or lost altogether. Still, this will likely be dismissed as an unavoidable accident and not be allowed to hamper humanity's space development—in the same way we grow hardened to the possibility of lives on Earth being lost in blizzards or by lightning strikes.

Yet, the time may come when we will know enough

about supernovas to be able to predict fairly accurately the chances of a nearby explosion taking place at a certain time. We may even be able to do intelligent solar-weather forecasting and predict the chances of strong solar flares. At such times, space would simply be cleared of unshielded personnel as far as possible, and people would wait until the worst of the danger is past before venturing out again.

The Next Supernova

If we are safely here on the Earth's surface, a supernova is not likely to be deadly, and if one appears in our own galaxy and is not hidden by dustclouds, it will form a glorious part of our night sky. A supernova that is moderately close will be much brighter than any star or planet in the sky and could (as did the Lupus supernova of 1006) vie in brightness with the Moon itself. And, of course, a bright supernova will even be visible in daylight for a period of time.

There hasn't been any supernova visible to the unaided eye, however, since 1604, and we have, in a sense, been cheated. For considering the rate at which supernovas form, we had every right to expect several to have blazed forth during the last 400 years.

If people have missed the chance of seeing a very bright, if temporary, pinpoint of light in the heavens, astronomers have missed considerably more than that. Were a bright supernova to burst into view, and were modern instruments focused upon it, we could find out in a few days more about supernovas, and about stellar evolution in general, than we have managed to learn during all the nearly four centuries since the last supernova was visible to the unaided eye.

How long must this celestial dearth continue? Is there any chance that we might see a bright supernova in the near future?

Yes, there is, We can even make reasonable guesses as to where it will appear.

In the first place, if a supernova flashes forth sometime during the next few years, it must be in its last stages before collapse right now. That means it must be a red giant. To be a spectacular sight when it explodes, it should be relatively close to us. Therefore, in considering candidates for the next supernova, we ought to concentrate, first, on the nearby red giants.

The nearest red giant to ourselves is Scheat, in the constellation of Pegasus. It is only about fifty parsecs away, but its diameter is about 110 times that of the Sun. This is small, as red giants go, and if this is as big as it is going to get, it is probably no more massive than the Sun and will not ever be a supernova. If it is still expanding, it has a considerable way to go before exploding and we need not expect a supernova, if so, for a million years or more.

Mira, or Omicron Ceti, is seventy parsecs away, but it has a diameter 420 times that of the Sun and is definitely more massive than the Sun. It is also pulsating irregularly, which may be a sign that it is in its final stages and is becoming increasingly unstable. It is a possible candidate for the next supernova, as seen from ringside Earth.

There are three comparatively nearby red giants, only about 150 parsecs away, that are each even more massive than Mira. One of these is Ras Algethi, in Hercules, with a diameter 500 times that of the Sun, and another is Antares, in Scorpio, with a diameter 640 times that of the Sun. Larger still is Betelgeuse, in Orion, which, like Mira, is pulsating. It is anywhere from fifteen to thirty times as massive as the Sun.

Betelgeuse, in fact, seems to be pre-supernova in a number of ways. It has an enormous stellar wind and is blowing off an amount of mass equal to $1/100,000$ the mass of our Sun each year. Another way of putting it is that it is losing material equal to the mass of the Moon every day and a half.

With such an enormous stellar wind, it is not surprising that Betelgeuse is surrounded by a shell of gas, which, according to recent studies, is abnormally low in carbon nuclei. Such a shortage of carbon is thought to go along with a high content of nitrogen nuclei, and some supernova remnants are found to be high in nitrogen. If, then, the outskirts of a red giant prove to be high in nitrogen, that would seem a clear sign that a supernova explosion cannot be far away.

To say that an astronomic event "cannot be far away" does not mean, however, that you ought to be watching the sky expectantly every night. In the lifetime of a star, "soon" may well mean a thousand or even ten thousand years. Betelgeuse may explode tomorrow (or it may have exploded nearly five hundred years ago, and the light of the explosion may finally reach us tomorrow), or it may not explode for several thousand years. We can't be sure.

Of course, if astronomers could only see a nearby supernova, *any* nearby supernova, they may learn enough about the conditions of such explosions to make it possible to time the next occasion much more precisely.

Betelgeuse, when it explodes, may well prove to be far brighter than any other supernova that has appeared during humanity's existence on Earth, since it is closer than any of the earlier ones. Betelgeuse is less than a tenth the distance of the great supernova of 1054, for instance.

Supernova Betelgeuse might, at its peak, rival the

light-intensity of the full Moon. However, while the full Moon spreads its light over a sizable circle so that no star-sized portion of it is intensely bright and it may therefore be looked at safely for as long as one wishes, Supernova Betelgeuse would pack all its light into a tiny point. It would not be wise, in that case, to stare at it in too much of a prolonged trance, for it might produce damage to the retina.

Supernova Betelgeuse, particularly if it exploded when our magnetic field was near zero, might produce a large enough flood of cosmic rays to bring about a noticeable increase in the genetic load of various organisms and might even lead to some extinctions. If it happened to explode when humanity was moving off the surface of Earth, but had not yet managed to construct adequate shielding for various structures being built, it might do serious damage to people in space. But, at the moment, there's nothing we can do about it.

It may be that Betelgeuse will not, after all, be the next star to produce a visible supernova. Some astronomers are convinced that the best candidate is Eta Carina, which, as I mentioned earlier, was first studied by John Herschel.

Eta Carina has an even stronger stellar wind than Betelgeuse, and it therefore has a denser shell of gas about it. This shell of gas absorbs some of the light that Eta Carina emits and makes it seem dimmer than it otherwise would appear. The shell releases the light eventually in the less energetic form of infrared radiation. The total energy, however, can't decrease, so that the infrared radiation has to be very great in quantity to make up for the lowered energy of each quantum. In fact, more infrared radiation reaches us from Eta Carina than from any other object in the sky that lies outside the solar system.

What's more, the shell of gas is low in carbon and high in nitrogen. Finally, Eta Carina is even more unstable than Betelgeuse and, in the past, has undergone comparatively minor explosions that nevertheless succeeded in making it, for a time at least, the second brightest star in the sky. It was then surpassed only by Sirius.

Sirius, however, is 2.7 parsecs away from us, while Eta Carina is 2,750 parsecs away. Considering that Eta Carina is a thousand times as far from us as Sirius is, for it to rival Sirius in brightness means that for a period of time it must have been nearly a million times as luminous as Sirius.

Eta Carina, then, may be closer to the edge than Betelgeuse, but, if Eta Carina explodes, it won't be as good a show. Eta Carina is nearly twenty times as far away as Betelgeuse, so that Supernova Eta Carina would appear only a little more than 1/400 as bright as Supernova Betelgeuse would. What's more, Eta Carina is located far down in the southern sky, so that when it does explode the result won't be visible in Europe or in most of the United States.

But then, Supernova Eta Carina would have less capacity to do damage than Supernova Betelgeuse would, and there's that to consider, too.

You see, then, that we have come a long way from Aristotle's vision of a calm, unchanging sky. We now know that it is a violent sky with enormously energetic events taking place here and there. We know that every once in a while we may witness, with the unaided eye, an energetic event such as the explosion of a star, and that such an event may not be entirely without danger to us here on Earth.

But we should rejoice and never complain. Our Sun wouldn't be what it is without the explosion and death of other suns; the Earth wouldn't exist in its present form, either; and we, and all our fellow life forms, wouldn't be here to enjoy our planet, our Sun, and, in the special case of people (including book readers), the sense of wonder that wells up within us any evening that we gaze at our galaxy strewn across the night sky.

INDEX

Flare, solar, 163, 164
  cosmic rays and, 243
  spaceflight and, 259
Flare stars, 170
Flemming, Walther, 219
Fossils, 203ff.
  evolution and, 213
Fox, Sidney Walter, 211
Fraunhofer, Joseph, 52
Friedman, Alexander Alexan-
  drovich, 142
Friedman, Herbert, 112

Galaxies, clusters of, 96, 143
  formation of, 147, 181, 182
  recession of, 140, 141
Galaxy (Milky Way), 76
  number of stars in, 47
Galileo, 28-31
Gamma rays, 238
Gamow, George, 143
Ganymede, 200
Gas clouds, interstellar, 182ff.
Gas giants, 196
Genes, 221
Genetic load, 248
Genetics, 221
Gilbert, William, 249
Globular clusters, 80
Gold, Thomas, 120
Goodricke, John, 33, 58
Grand Unified Theories, 146
Great Dyings, 253ff.
Guest stars, 14ff.
Gum, Colin S., 105
Gum nebula, 105
  pulsar in, 119
GUTS, 146

Hale, George Ellery, 163
Halley, Edmond, 35, 36
Halley's Comet, 35
Harkins, William Draper, 54
Hartmann, Johannes Franz,
  186

Hartwig, Ernst, 84
Hazard, Cyril, 104
HD-226868, 135
Heliocentric theory, 19
Helium, 150
  sun and, 55
  universe and, 179, 180
Helium-3, 155-57
Helium-4, 156ff.
Helium-5, 158
Helium fusion, 65, 66
Helmholtz, Hermann L. F.
  von, 50, 51
Henderson, Thomas, 38
Hercules, 261
  nova in, 43
Herschel, John, 38
Herschel, William, 58, 78, 80
Hertzsprung, Ejnar, 63
Hess, Victor Franz, 237
Hevelius, Johannes, 32
Hewish, Antony, 117
Hipparchus, 8ff.
Holward of Franekar, 31
Horseshoe crab, 204
Hubble, Edwin Powell, 92ff.,
  101, 141
Hubble's law, 141ff.
Huggins, William, 84, 139
Humason, Milton La Salle, 141
Huygens, Christian, 78
Hydrogen, atomic structure
  of, 149
  isotopes of, 154, 155
  sun and, 54ff.
  universe and, 179, 180
Hydrogen-1, 155
Hydrogen-2, 155
Hydrogen-3, 155, 156
Hydrogen fusion, 54, 55, 160

Ices, 197
Inflationary Universe, 146
Inheritance, Mendelian laws
  of, 217, 218

270